Troubleshooting and Repairing Computer Monitors

Troubleshooting and Repairing Computer Monitors

2nd Edition

Stephen J. Bigelow

McGraw-Hill

New York   San Francisco   Washington, D.C.   Auckland   Bogotá   Caracas   Lisbon   London
Madrid   Mexico City   Milan   Montreal   New Delhi   San Juan   Singapore   Sydney   Tokyo   Toronto

Library of Congress Cataloging-in-Publication Data

Bigelow, Stephen J.
    Troubleshooting and repairing computer monitors / Stephen J.
Bigelow.—2nd ed.
       p.   cm.
    Includes index.
    ISBN 0-07-005733-8 (hc).—ISBN 0-07-005734-6 (pc)
    1. Computer terminals—Maintenance and repair. I. Title.
TK7887.8.T4B53 1997
621.39'87—dc20   96-34038
                                      CIP

McGraw-Hill

A Division of The McGraw·Hill Companies

1 2 3 4 5 6 7 8 9 0 DOC/DOC 9 0 1 0 9 8 7 6

ISBN 0-07-005734-6 (PBK)

ISBN 0-07-005733-8 (HC)

*The sponsoring editor for this book was Steve Chapman, the editing supervisor was
Fred Bernardi, and the production supervisor was Pamela A. Pelton. It was set in
ITC Century Light by North Market Street Graphics.*

Printed and bound by R. R. Donnelley & Sons Company.

The following trademarks and servicemarks may appear in this book. Every effort has been
made to cover each appropriate reference. All other trademarks and servicemarks remain the
property of their respective owners: *IBM* is a trademark of International Business Machines;
MS-DOS, Windows, and *Windows 95* are trademarks of Microsoft Corporation; *Tandy* and
Radio Shack are trademarks of Tandy Corporation; *The PC Toolbox* is a trademark of
Dynamic Learning Systems; *i86, i286, i386, i486, Pentium,* and *Intel* are trademarks of
Intel Corporation; *CompuServe* is a trademark of CompuServe Incorporated; *MultiSync*
is a trademark of NEC Technologies, Inc.

McGraw-Hill books are available at special quantity discounts to use as premiums and
sales promotions, or for use in corporate training programs. For more information,
please write to the Director of Special Sales, McGraw-Hill, 11 West 19th Street,
New York, NY 10011. Or contact your local bookstore.

This book is printed on recycled, acid-free paper containing a minimum of
50 percent recycled, de-inked fiber.

THIS BOOK IS DEDICATED TO MY WONDERFUL WIFE, KATHLEEN.
Without her loving support and encouragement, this book would
still have been possible, but not nearly worth the effort.

Disclaimer and cautions

It is IMPORTANT that you read and understand the following information. Please read it carefully!

The repair of computer monitors involves some amount of personal risk. Use **extreme** caution when working with ac and high-voltage power sources. Every reasonable effort has been made to identify and reduce areas of personal risk. You are instructed to read this book carefully *before* attempting the procedures discussed. If you are uncomfortable following the procedures that are outlined in this book, **do *not* attempt them**. Refer your service to qualified service personnel.

Neither the author, the publisher, nor anyone directly or indirectly connected with the publication of this book shall make any warranty, either expressed or implied, with regard to this material, including, but not limited to, the implied warranties of quality, merchantability, and fitness for any particular purpose. Further, neither the author, publisher, nor anyone directly or indirectly connected with the publication of this book shall be liable for errors or omissions contained herein or for incidental or consequential damages, injuries, or financial or material losses resulting from the use, or inability to use, the material contained herein. This material is provided *as is*, and the reader bears all responsibilities connected with its use.

The products, materials, equipment, manufacturers, service providers, and distributors listed and presented in this book are shown for example purposes only. Their use in this book does not constitute an endorsement by the author, publisher, or anyone directly or indirectly connected with the publication of this book.

Contents

Preface

Monitors have played a vital role in computers long before the computer ever became "personal." Faster and less cumbersome than teletype terminals or printers, monitors helped to make computers more "user friendly." As computers advanced and became more available, monitors proliferated as well. The bland, monochrome text displays of years past have given way to monitors offering photorealistic color and resolution as well as energy efficiency. Today, the PC monitor is a *necessary* part of a computer rather than just a peripheral. With literally millions of monitors in operation around the world, the ability to test and service those monitors is an important consideration for all PC users and electronics enthusiasts. This book is designed to explain the operations of a PC video system, show you how monochrome and color CRTs work, and provide you with service information that you can use to repair your monitor and keep it running longer.

Book highlights and the second edition

Chapter 1 presents an overview of monitor concepts, terminology, and specifications. The second edition offers additional data on monochrome phosphors and the CIE Chromaticity Chart. Chapter 2 continues the discussion with an overview of the monitor's internal assemblies and operations, with a special emphasis on energy-saving systems and CRT troubleshooting. A selection of test instruments and tools is discussed in Chapter 3, with a new section dedicated to computer-aided troubleshooting tools. These first three chapters are designed for the beginner, although experienced troubleshooters may find them to be a helpful review.

Chapter 4 presents a thorough overview of video adapters—the link between a PC and monitor. Readers will learn about conventional video adapters as well as graphics accelerators. The second edition includes details of the video feature connector (VFC). A variety of symptoms and troubleshooting tests will help readers identify and correct video adapter problems quickly and efficiently. Chapter 5 covers a broad cross-section of monitor alignment procedures that can be used to keep a monitor (especially a color monitor) running at top performance. This edition includes a brief discussion of the difficult "dynamic convergence" alignment process. The companion software available for this book is used extensively in the alignment process.

Chapters 6, 7, and 8 detail typical monitors to the component level. Power supplies are discussed in Chapter 6. Linear, switching, and high-voltage generation circuits are presented along with step-by-step troubleshooting procedures. Liquid crystal display (LCD) backlight power supplies are also covered. Chapter 7 deals with the classical monochrome monitor circuit, explaining the video, vertical, horizontal, and flyback circuits at the component level. An array of practical symptoms and solutions guide the reader through each procedure. Chapter 8 concentrates on color monitors, presenting the actual component-level video, vertical, horizontal, and flyback circuits for discussion. Additional troubleshooting procedures indigenous to color monitors lend deeper perspective to the chapter. Both component-level and subassembly-level repair solutions are presented, so all levels of readers can take definitive repair actions. The second edition of the book provides a large number of new symptoms that will expand your understanding of monitor troubleshooting.

No book on PC display systems would be complete without a review of flat-panel displays such as LCDs and plasma panels. Chapter 9 presents a review of flat-panel technologies and assemblies and then shows the reader how to identify and correct problems with LCD and plasma displays. Chapter 10 reviews the companion software for this book (updated for MONITORS version 2.01). A form is included for easy ordering.

Every possible measure was taken to ensure a thorough and comprehensive book. Your comments, questions, and suggestions about this book are welcome at any time, as are any personal troubleshooting experiences that you may like to share. Feel free to write to me directly or contact me through e-mail:

Stephen J. Bigelow
Dynamic Learning Systems
P.O. Box 282
Jefferson, MA 01522-0282 USA
CompuServe: 73652,3205
Internet: sbigelow@cerfnet.com
WWW: http://www.dlspubs.com/

Acknowledgments

IN TODAY'S FAST-MOVING WORLD OF PERSONAL COMPUTERS, it is virtually impossible to prepare a thorough, complete book on computer service without the cooperation and support of other individuals and corporations. I offer my heartfelt thanks for the material and technical support that helped to make this book:

☐ Gregg Elmore, Manager of Marketing Communications, B+K Precision

☐ Scott Evans, Advertising and Sales Promotion Manager, NEC Technologies, Inc.

☐ Brad Johnson, Merchandising Manager, Sencore, Inc.

☐ Janice B. Schopper, Marketing Communications Coordinator, Sharp Electronics Corporation

☐ Mary Pat Strouse, Marketing Communications Assistant, Network Technologies, Inc.

☐ Special thanks to Ron Trumbla, Media Relations Representative at Tandy Corporation. The extensive reprints permitted from Tandy service literature have greatly enhanced the book.

And of course, I wish to thank Roland S. Phelps and the development staff at McGraw-Hill for their endless patience and consideration in making this book a reality.

Contemporary computer monitors

THE COMPUTER MONITOR HAS EVOLVED FAR BEYOND THE simple output device that replaced old-fashioned teletype terminals (Fig. 1-1). From humble beginnings as basic monochrome text displays, the monitor has grown to provide real-time photorealistic images of unprecedented quality and color. Monitors have allowed stunning graphics and information-filled illustrations to replace the generic "command line" user interface of just a few years ago. Monitors have become our *virtual window* into the modern computer. The explosive growth of portable computers (i.e., notebook, subnotebook, and palmtop systems) has also fueled the need to understand and service both monochrome and color liquid crystal flat-panel displays (LCDs). As advances in PC technology bring computers, telephones, televisions, and multimedia closer together (and environmental concerns demand everless power consumption), monitors and liquid crystal displays will continue to be a focal point of computer development.

With many millions of computers now in service, the economical maintenance and repair of computer monitors represent serious challenges to technicians and enthusiasts alike. Fortunately, the basic principles and operations of a computer monitor have changed very little since the days of "terminal displays." Before any discussion of monitor repair can begin, however, you need to understand the characteristics and specifications that define a monitor and how that monitor functions.

Understanding monitor specifications

While PCs are defined by fairly simple and well-understood specifications such as RAM size, hard drive space, and clock speed, monitor specifications describe a whole series of physical properties that PCs never deal with. With this in mind, perhaps the best

■ 1-1 *An NEC MultiSync 3FGe monitor. (NEC Technologies, Inc.)*

introduction to monitor technology is to discuss each specification in detail and show you how each specification and characteristic affects a monitor's performance.

CRT

The *cathode-ray tube* (or CRT) is essentially a large vacuum tube. One end of the CRT is formed as a long, narrow neck, while the other end is a broad, almost flat surface (Fig. 1-2). A phosphor coating is applied inside the CRT along the front face. The neck end of the CRT contains an element (called the *cathode*) which is energized and heated to very high temperatures (much like an incandescent lamp). At high temperatures, the cathode liberates electrons. When a very high positive voltage potential is applied at the front face of the CRT, electrons liberated by the cathode (which are negatively charged) are accelerated toward the front face. When the fast-moving electron strikes the phosphor on the front face, light is produced. By directing the stream of electrons across the front face, a visible image is produced. Of course, there are other elements needed to control and direct the electron stream, but this is CRT operation in a nutshell.

CRT face size (or *screen size*) is generally measured as a diagonal dimension—that is, a 43.2-cm (17-in.) CRT is 43.2 cm (17 in.) between opposing corners. Larger CRTs are more expensive, but

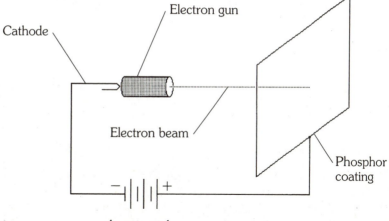

Cathode

Electron gun

Electron beam

Phosphor coating

dc potential

■ **1-2** *Simplified diagram of a conventional monochrome CRT.*

produce larger images which are usually easier on the eyes. A CRT specification will also list the type of phosphor(s) used (e.g., P22) in addition to antiglare (or any special finish) coatings employed during manufacture.

Pixels and resolution

The picture element (or *pixel*) is the very smallest point that can be controlled on a CRT or LCD. For monochrome displays, a pixel may simply be turned on or off. For a color display, a pixel may assume any of a number of different colors. Pixels are combined in the form of an array (rows and columns). It is the size of that pixel array which defines the display's *resolution*. Thus, resolution is the total number of pixels in width by the total number of pixels in height. For example, a typical enhanced graphics adapter (EGA) resolution is 640 pixels wide by 350 pixels high, while a standard video graphics array (VGA) resolution is 640 pixels wide by 480 pixels high. Typical super VGA (SVGA) resolution is 800 pixels wide by 600 pixels high and higher. Resolution is important for computer monitors since higher resolutions allow finer image detail. Table 1-1 offers a brief index of monitor resolutions.

Triads and dot pitch

Monochrome CRTs use a single, uniform phosphor coating (usually white, amber, or green), whereas color CRTs use three color phosphors (red, green, and blue) arranged as triangles (or *triads*). Figure 1-3 illustrates a series of color phosphor triads. On a

	Resolution	
Monitor type	**Height**	**Width**
Color Graphics Adapter (CGA)	320 pixels	200 pixels
Enhanced Graphics Adapter (EGA)	640 pixels	350 pixels
Video Graphics Array (VGA)	640 pixels	480 pixels
Super VGA (SVGA)	800 pixels	600 pixels
	1024 pixels	768 pixels

color monitor, each triad represents one pixel (even though there are three *dots* in the pixel). By using the electron streams from three electron guns—one gun for red, one for blue, and another for green—to excite each dot, a broad spectrum of colors can be produced. The three dots are placed so close together that they appear as a single point to the unaided eye.

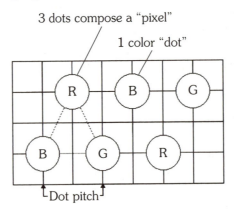

■ **1-3** *Color phosphors and dot pitch.*

The quality of a color image is related to just how close each of the three dots are to one another. The closer together they are, the purer the image appears. As the dots of a pixel are spaced further apart, the image quality degrades because the eye can begin to discern the individual dots in each pixel. This results in lines that no longer appear straight, and colors are no longer pure. *Dot pitch* is a measure of the distance between any two phosphor dots in a pixel. Displays with a dot pitch of 0.31 mm or less generally provide adequate image quality.

Shadow and slot masks

The *shadow mask* is a thin sheet of perforated metal that is placed in the color CRT just behind the phosphor coating. Elec-

4

tron beams from each of the three electron guns are focused to converge at each hole in the mask, *not* at the phosphor screen (Fig. 1-4). The microscopic holes act as apertures that let the electron beams through *only* to their corresponding color phosphors. In this way, any stray electrons are *masked,* and color is kept pure. Without a mask, stray electrons from one electron beam may accidentally excite nearby color phosphors resulting in unwanted color combinations. Some CRT designs substitute a shadow mask with a *slot mask* (or *aperture grille*) which is made up of vertical wires behind the phosphor screen. Dot pitch for CRTs with slot masks is defined as the distance between each slot. Keep in mind that monochrome CRT designs do not use a mask because the entire phosphor surface is the same color.

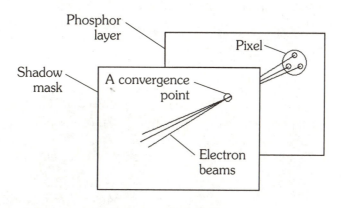

* Sizes and distances are NOT shown to scale

■ **1-4** *The action of a shadow mask.*

Convergence

Remember that *three* electron guns are used in a color monitor, and each gun excites a particular color phosphor. All three electron beams are tracking around the screen simultaneously, and the beams converge at holes in the shadow mask. This *convergence* of electron beams is closely related to color purity in the screen image. Ideally, the three beams converge perfectly at all points on the mask, and the resulting color is perfectly pure throughout (i.e., pure white). If one or more beams do not converge properly, however, the image color will not be pure. In most cases, poor convergence will result in colored shadows. For example, you may see a red, green, or blue shadow when looking at a white line. Serious convergence problems can result in a blurred or distorted image. Monitor specifications usually list typical convergence error as *misconvergence* at both the display center and the overall display area. Typical center misconvergence runs approximately 0.45 mm, and overall display area misconvergence is about 0.65 mm. Larger numbers result in poorer convergence.

Pincushion and barrel distortion

The front face of most CRTs is slightly convex (bulging outward). On the other hand, images are perfectly square. When a square image is projected onto a curved surface, distortion results. Ideally, a monitor's raster circuits will compensate for this screen shape so that the image appears square when viewed at normal distances. In actual practice, however, the image is rarely square. The sides of the image (top to bottom and left to right) may be bent slightly inward or slightly outward. Figure 1-5 illustrates an exaggerated view of these effects. *Pincushioning* occurs when sides are bent inward making the image's boarder appear concave. Other straight lines within the image may also appear to curve toward the image center. *Barreling* occurs when the sides are bent outward making the image's boarder appear convex. Straight lines in the image may appear to bend outward toward the edges of the screen. In a properly aligned monitor, these distortions should be just barely noticeable (no more than 2.0 or 3.0 mm). Keep in mind that many technical people refer to barrel distortion as pincushioning as well.

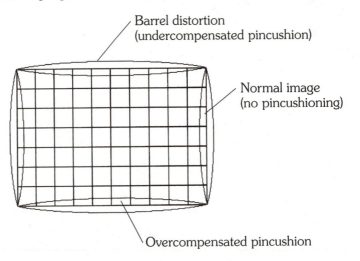

Barrel distortion
(undercompensated pincushion)

Normal image
(no pincushioning)

Overcompensated pincushion

■ **1-5** *The effects of pincushion distortion on monitor images.*

Horizontal scanning, vertical scanning, raster, and retrace

To understand what *scanning* is, you must first understand how a monitor's image is formed. A monitor's image is generated one horizontal line of pixels at a time starting from the upper left corner of the display (Fig. 1-6). As the beams travel across the line, each pixel is excited based on the video data contained in the corresponding location of video RAM on the video adapter board. When a line is complete, the beam turns off (*horizontal* blank) and is directed horizontally (and slightly vertically) to the beginning of the next line. A new horizontal line can then be drawn. This process continues until all horizontal lines are drawn and the beam is in the lower right corner of the display. When the image is complete, the beam turns off (*vertical* blank) and is redirected to the upper left corner of the display to start all over again. The rate at which horizontal lines are drawn is known as the *horizontal scanning rate* (sometimes called *horizontal sync*). The rate at which a complete "page" of horizontal lines is generated is known as the *vertical scanning rate (vertical sync)*. Both the horizontal and vertical blanking times are known as *retrace times* because the deactivated beams are "retracing" their path before starting a new trace. A typical horizontal retrace time is 5 ms, and the typical vertical retrace time is 700 ms (though these times will vary depending on the screen resolution being used). This continuous horizontal and vertical scanning action is known as *raster.*

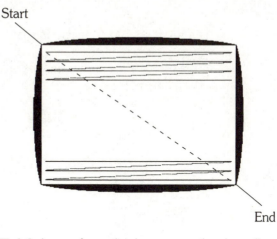
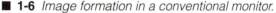

Start

End

■ **1-6** *Image formation in a conventional monitor.*

We can easily apply numbers to scanning rates to give you an even better idea of their relationship. A typical VGA monitor with a resolution of 640 × 480 pixels uses a horizontal scanning rate of 31.5 kHz. This means that 31,500 lines can be drawn in 1 s, or a single line of 640 pixels can be drawn in 31.7 µs. As there are 480 horizontal lines to be drawn in one page, a complete page can be drawn in 15.2 ms (480 × 31.7 µs). If a single page can be drawn in 15.2 ms, the screen can be refreshed 65.7 times per second (65.7 Hz); this is roughly the vertical rate that will be set for VGA operation at 640 × 480 resolution. In actual practice, the vertical scanning rate will be set to a whole number such as 60 Hz, which leaves a lot of spare time for blanking and synchronization. It was discovered early in TV design that vertical scanning rates under 60 Hz resulted in perceivable flicker that causes eye strain and fatigue. You can start to see now that horizontal scanning rates are *not* chosen arbitrarily. The objective is to select a horizontal frequency that will cover a page's worth of horizontal pixel lines for any given resolution at *about* 60 times per second. Newer monitor designs are now using vertical scanning rates of 72 to 80 Hz.

Let's look at an older monitor and see if this concept holds true. A color graphics adapter (CGA) monitor offers a resolution of 320 × 200 pixels, and the standard horizontal scan rate for this resolution is 15.6 kHz. If 15,600 horizontal lines can be drawn in 1 s, a single horizontal line can be drawn in 64.1 μs. At this speed, a page of 200 horizontal lines can be draw in 12.8 ms (200 × 64.1 μs), and the image can be refreshed 78 times in 1 s (78 Hz). The actual vertical scanning rate can easily be dropped to 60 Hz which allows plenty of time for older display circuitry to keep up.

Now let's look at an SVGA monitor working at a resolution of 800 × 600 pixels. Ideally, we need a horizontal scanning rate that will allow the display to refresh 600 horizontal lines about 60 times per second. The standard horizontal scanning rate at 800 × 600 is 38 kHz, which will draw a single horizontal line in 26.3 μs. A page of 600 lines can be drawn in 15.8 ms (600 × 26.3 μs). At that familiar rate, the display can be refreshed 63.4 times in 1 s (63.4 Hz). As with other designs, the actual vertical refresh rate can be set to 60 Hz, which provides extra time for synchronization. Table 1-2 compares the scanning rates for current monitor resolutions.

Interlacing

Table 1-2 introduces another important concept of horizontal scanning known as *interlacing*. Images are "painted" onto a display one horizontal row at a time, but the sequence in which those lines are drawn can be noninterlaced or interlaced. As you see in Fig.1-7, a *noninterlaced* monitor draws all of the lines that comprise an image in one pass. This is preferable since a noninterlaced image is easier on your eyes—the entire image is refreshed at the vertical scanning frequency—so a 60-Hz vertical scanning rate will update the entire image 60 times in 1 s. An *in-*

■ Table 1-2 Scan rates versus monitor resolution

Monitor	Resolution	Horizontal scan	Vertical scan
MDA	720 × 348	18.43 kHz	50.0 Hz
CGA	320 × 200	15.85 kHz	60.5 Hz
EGA	640 × 350	21.80 kHz	60.0 Hz
VGA	640 × 350	31.50 kHz	70.1 Hz alternate config.
VGA	640 × 480	31.47 kHz	60.0 Hz
VGA	640 × 480	37.90 kHz	72.0 Hz VESA config.
SVGA	800 × 600	38.00 kHz	60.0 Hz
SVGA	800 × 600	35.16 kHz	56.0 Hz
SVGA	800 × 600	37.60 kHz	72.0 Hz
SVGA	1024 × 768	35.52 kHz	87.0 Hz interlaced (8514A)
SVGA	1024 × 768	48.80 kHz	60.0 Hz Sony config.
SVGA	1024 × 864	54.00 kHz	60.0 Hz DEC config.
SVGA	1006 × 1048	62.80 kHz	59.8 Hz Samsung config.
SVGA	1280 × 1024	70.70 kHz	66.5 Hz DEC config.
SVGA	1600 × 1280	89.20 kHz	66.9 Hz Sun config.

terlaced display draws an image in two passes. Once the first pass is complete, a second pass fills in the rest of the image. The *effective* image refresh rate is only *half* the stated vertical scanning rate. The typical 1024 × 768 SVGA monitor of Table 1-2 shows a vertical scanning rate of 87 Hz, but because the monitor is interlaced, effective refresh is only *43.5 Hz*. Screen flicker is much more noticeable.

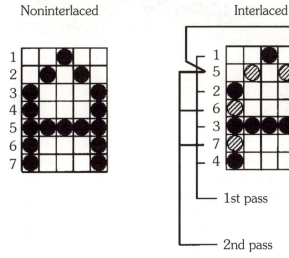

■ **1-7** *Interlaced versus noninterlaced scanning.*

Bandwidth

In the very simplest terms, the *bandwidth* of a monitor is the *absolute maximum* rate at which pixels can be written to the display. Typical VGA displays offer a bandwidth of 30 MHz. That is, the monitor could generate up to 30 million pixels per second on the display. Consider that each scan line of a VGA display uses 640 pixels and the horizontal scan rate of 31.45 kHz allows 31,450 scan lines per second to be written. At that rate, the monitor is processing 20,128,000 pixels per second (640 pixels per scan line × 31,450 scan lines per second), well within the monitor's 30 MHz bandwidth. The very newest high-resolution color monitors offer bandwidths of 135 MHz. Such high-resolution 1280 × 1024 monitors with scanning rates of 79 kHz would need to process at least 101,120,000 pixels per second (101.12 MHz) (1280 pixels per scan line × 79,000 scan lines per second), so enhanced bandwidth is truly a necessity. Figure 1-8 provides a comparison of typical monitor bandwidths.

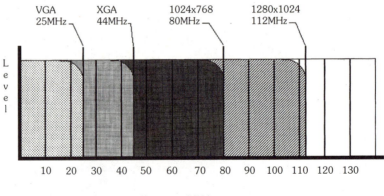

■ **1-8** *Typical bandwidth requirements for video standards.*

Swim, jitter, and drift

The electron beam(s) that form an image are directed around a display using variable magnetic fields generated by separate vertical and horizontal *deflection coils* mounted at the CRT's neck. The analog signals that drive each deflection coil are produced by horizontal and vertical deflection circuitry (which this book will detail in later chapters). Ideally, deflection circuitry should steer the electron beam(s) precisely the same way in each pass. This would result in an absolutely rock-solid image on the display. In

10

the real world, however, there are minute variations in the placement of images over any given period of time. *Jitter* is a term applied to measuring such variation over a 15-s period. *Swim* (sometimes called *wave*) is a measure of position variation over a 30-s period. *Drift* is a measure of position variation over a 1-min period. Note that all three terms represent essentially the same problem, but over different amounts of time. Swim, jitter, and drift may be expressed as fractions of a pixel or as physical measurements such as millimeters.

Brightness

When an electron beam strikes phosphor, light is liberated. The *brightness* of an image indicates how much light is generated when an image is formed. Brightness is measured in foot lumens (ft lm). While the actual physics of foot lumens and brightness are beyond the scope of this book, simply consider that most monitors provide visible brightness levels of 50 to 60 ft lm depending on the CRT's brightness setting and how much white is contained in the overall image. Larger numbers indicate brighter displays, whereas smaller numbers suggest dimmer displays (i.e., background raster is usually about 0.3 ft lm). Keep in mind that brightness is typically measured while a pure white square is shown in the center 20 percent of the display. Not all monitors provide a specification for brightness since it is a difficult and rather subjective quantity to measure. Precise scientific instrumentation is needed for an accurate measurement.

Centering

An image should appear centered in the display (with all horizontal and vertical centering controls at their midrange positions). Centering specifications indicate how close the image center actually comes to the screen center. Centering in the horizontal and vertical orientations should generally be within ±7 mm.

Video signal

This specification lists signal levels and characteristics of the analog video input channel(s). In most cases, a video signal in the $0.7 V_{pp}$ (peak to peak) range is used. Circuitry inside the monitor amplifies and manipulates these relatively small signals. A related specification is input impedance which is often at 75 Ω. Older monitors using digital (on/off) video signals typically operate up to 1.5 V.

Synchronization and polarity

After a line is drawn on the display, the electron beams are turned off (blanked) and repositioned to start the next horizontal line. However, no data are contained in the retrace line. For the new line to be "in sync" with the data for that line, a *synchronization* pulse is sent from the video adapter to the monitor. There is a separate pulse for horizontal synchronization and vertical synchronization. In most current monitors, synchronization signals are edge triggered TTL (transistor-transistor logic) signals. *Polarity* refers to the edge that triggers the synchronization. A falling trigger (marked "−" or "positive/negative") indicates that synchronization takes place at the high-to-low transition of the sync signal. A leading trigger (marked "+" or "negative/positive") indicates that synchronization takes place on the low-to-high transition of the sync signal. Table 1-3 shows the synchronization polarity for four common resolutions.

■ **Table 1-3 Synchronization polarities versus resolution**

	Number of horizontal lines			
	350	400	480	768
H sync polarity	+	−	−	+
V sync polarity	−	+	−	+

Power

Power specifications include the ac voltage (measured in volts) and current (in amps) requirements of the monitor as well as the resulting power dissipation (watts). Power frequency (hertz) is also usually listed. Current domestic (U.S.) monitors generally require 90–132 V ac at 47–63 Hz, draw 3 A, and dissipate between 70–90 W. With increasing concern about the environment, monitor manufacturers are beginning to develop "green" monitors that will power down to some low level on demand or after a user-definable idle period.

Environment

A set of environmental specifications outline the conditions under which a monitor can safely operate. Operating temperature, storage temperature, noncondensing humidity range, and operating altitude are the four typical environmental ratings that you should be concerned with. An average monitor works between 0 to 40°C in an environment with 10 to 90 percent noncondensing humidity, but the monitor can be stored in temperatures between −30 and 60°C. The average monitor will work below 3079 m (10,000 ft) in altitude.

Evaluating a new monitor

You will eventually buy a new monitor or recommend a new monitor to someone else. However, all monitors are *not* created equal. Variations in CRT phosphors, video drive circuits, interlacing schemes, and so on simply make some monitors better than others. The following guidelines may help you make the best choice for you or your customer.

1. *Never buy a monitor without seeing it for yourself.* No matter how nice a monitor image appears in an advertisement, you should see the unit in actual operation first.

2. *Check the focus and high-voltage regulation.* With a square image displayed (preferably a bright white square), turn the monitor brightness up and down. If the edges of the image bloom or swell, the high-voltage regulation system may be weak. If there is text or other detail in the image, it may lose focus at high brightness levels.

3. *Check the convergence.* Display an image such as a Windows desktop and inspect the colors around each icon or boarder. If colors appear to "bleed" from around text or icons, the monitor's convergence may be off. This will cause mild distortion in the image, which can add to eye strain and fatigue.

4. *Check the aspect ratio.* Display a circle and make sure that the circle appears round. If not, the monitor's mode detection circuitry may not be set correctly.

5. *Check the flicker.* Your direct line of sight (the "conscious" eye) tends to filter out monitor flicker, but your peripheral vision does not. That is why eye strain and fatigue due to flicker are such problems. Just because your eyes do not register it, doesn't mean it's not there. When observing a monitor, avert your eyes to something near the monitor, and you will see the flicker in your peripheral vision. If the flicker is noticeable, pass the monitor by.

Understanding video signals

A PC video system is composed of two items: the video adapter board and the monitor itself. The *video adapter* is actually the heart of a video system. Video commands and data from the com-

13

puter are passed to the adapter board via its expansion bus (the card edge connector where the board plugs into the motherboard). The video board converts PC data into a graphic image that is stored in on-board video memory (called *video RAM* or *VRAM*). The video board then translates the graphic image into a series of color and synchronization signals that correspond to the video mode in use (e.g., CGA, EGA, VGA, etc.). Color and synchronization signals are then passed to a standard connector as shown in Fig. 1-9. Current VGA-compatible video boards use a 9-pin or 15-pin subminiature D-shell connector. The pinout of each pin is given in Table 1-4.

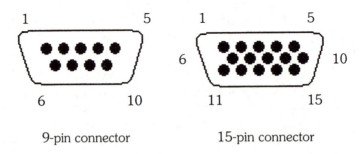

9-pin connector   15-pin connector

■ **1-9** *Video connector pinouts.*

■ **Table 1-4 Video connector pinouts**

	TTL			Analog	
Pin	8	16	64	9 pin	15 pin
1	Gnd	Gnd	Gnd	Red	Red
2	n/c	n/c	Red Inten.	Green	Green
3	Red	Red	Red	Blue	Blue
4	Green	Green	Green	Hsync	Gnd
5	Blue	Blue	Blue	Vsync	Gnd
6	n/c	Intensity	Green Inten.	Gnd	Red Gnd
7	n/c	n/c	Blue Inten.	Gnd	Green Gnd
8	Hsync	Hsync	Hsync	Gnd	Blue Gnd
9	Vsync	Vsync	Vsync	Gnd	n/c
10	—	—	—	—	Gnd
11	—	—	—	—	Gnd
12	—	—	—	—	n/c
13	—	—	—	—	Hsync
14	—	—	—	—	Vsync
15	—	—	—	—	n/c

The monitor connects to the video adapter's subminiature D-connector. By comparison, the monitor's task is remarkably simple: Synchronize the sweep of the CRT's electron beam with the incoming video data to create a visual image. By understanding the nature of video signals, you can gain a lot of insight into monitor operation.

There are two types of video signals: TTL and analog. This refers to the way in which color signals are sent to the monitor, so you will often hear monitors categorized as *TTL* or *analog*. In a TTL monitor, color signals take the form of logic levels (1s and 0s). A monochrome TTL monitor uses one or two signal lines. When one TTL line is used, only black and a single monochrome color may be produced. In this case, the actual color would be the CRT's phosphor color such as amber (yellow), green, or white. If two signal lines are used in a monochrome TTL monitor, one line turns the pixel on or off while the other line controls high or low intensity. Two shades of the phosphor color can then be produced.

Early TTL color monitors used three color signals: one for red, green, and blue as shown in Fig. 1-10. Three signals could be combined to produce 2^3 (or 8) unique colors. You can see the video pinout for this 8-color TTL monitor in Table 1-4. The problem with this signal technique is its obvious color limitations. When the red signal is on, it is fully on and red is saturated. To make brown, red and green are both turned on fully. In order to improve the diversity of available colors (increase the palette), intensity controls

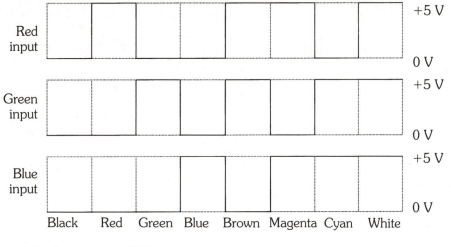

■ **1-10** *An example of TTL color signals.*

were added to modulate the saturation of each color. By adding a single master intensity line, the palette can be doubled from 8 to 16 colors. The jump from 16 to 64 colors employed a separate intensity control for each color as shown in Table 1-4.

In principle, TTL signaling works just fine. It could be argued that TTL signaling would work for any size of palette, but consider what happens when a 256-color palette is used: 5 intensity lines would be needed. Each time the palette is doubled, a new intensity line is added. For a 262,144 color palette, *37,768* intensity lines would be needed to modulate the original 8 colors. You can imagine how unwieldy it would be to try and connect such an arrangement. A more elegant solution was demanded as designers sought to expand the color palette beyond 64 colors. The solution was to replace TTL color and intensity signals with *analog* color saturation signals. Thus, the *analog* color monitor was developed.

Three analog signals are required: red, green, and blue. The amount of voltage corresponds directly to the saturation of the color. Analog voltages range from 0.0 to 0.7 V_{pp} as in Fig. 1-11. At 0 V, the color is off. If the red, green, and blue signals were all 0 V, the color would be black. At 0.7 V_{pp}, the color is saturated. If the red, green, and blue signals were all 0.7 V_{pp}, the resulting color would be pure white. Color is adjusted by varying the voltage level for each of the three primary colors. If the level of all three primary colors is equal, the resulting color will be a shade of gray as shown in Fig. 1-11.

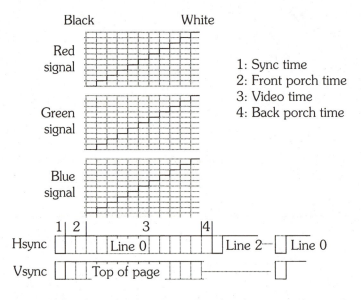

1: Sync time
2: Front porch time
3: Video time
4: Back porch time

■ **1-11** *An example of analog color signals.*

Notice that the voltages in Fig. 1-11 are not smooth as "true" analog signals would be, but vary in discrete steps. That is because the video adapter generates the analog signals using a series of digital-to-analog converters (DACs). Digital data that represent the saturation level of each primary color are converted to a corresponding voltage. This brings us back to the question of the palette: How many colors can be produced? The answer is in the number of steps each DAC is capable of. For example, if each of the three color DACs has 4 output levels, the total number of colors that can be produced is 64 ($4 \times 4 \times 4$). If each DAC has 8 steps, 512 ($8 \times 8 \times 8$) unique colors are possible. Modern DACs can resolve 256 steps, which opens the possibility of 16,777,216 (16.8 million) ($256 \times 256 \times 256$) colors, photorealistic color so subtle that the human eye cannot resolve the difference in shades.

Regardless of whether color data are supplied to a monitor in TTL or analog form, the data have to be synchronized to ensure that the right pixel shows up at the right place on the display. Synchronization is provided by two TTL signals: horizontal synchronization and vertical synchronization. A *horizontal synchronization* (Hsync) pulse is needed at the beginning of every horizontal scan line as illustrated in Fig. 1-12. If a VGA display has 480 lines and a horizontal scan rate of 31.45 kHz, an Hsync pulse would be needed every 31.7 μs. The duration of the logic 0 sync pulse itself is known as *sync time*. There is a small period after the sync pulse passes and before video data begin, which is the *back porch time*. After the line of video data is complete, there is a brief period before the next sync

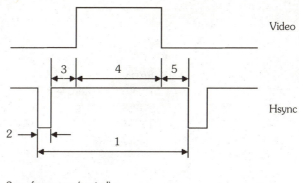

Video

Hsync

1 Sync frequency (period)
2 Sync time
3 Back porch time
4 Active video time
5 Front porch time

■ **1-12** *Video sync signal characteristics.*

pulse starts, which is the *front porch time.* A *vertical synchronization* (Vsync) pulse is needed at the beginning of every page. If the vertical refresh frequency is 60 Hz, a Vsync pulse is needed every 16.67 ms. Remember that both synchronization signals are provided by the video board based on the selected video mode.

Delivering the synchronization signals to the monitor presents another problem. There are three methods of carrying sync signals: separate, composite, and composite-on-video. *Separate* synchronization is just as the name implies. Hsync and Vsync signals are carried along individual signal lines as shown in Fig. 1-11. This is by far the most popular method of configuring an analog signal because the signals provide very reliable control over the image. Both the horizontal and vertical synchronization signals can be combined on the same line to produce a *composite* sync signal. It is also possible to ride the combined synchronization signals on one of the three video lines (usually the green video line). This is known as *composite-on-video* or *composite-on-green.* In general, composite signal techniques are not popular since additional monitor circuitry is needed to separate the signals, and precision (thus, video quality) is lost in the separation process. That is, the image tends to be less stable when attempting to separate composite sync signals.

18

Inside today's CRT displays

At this point, it is possible to take an initial look inside a typical CRT-based monitor and examine its essential subsections. Figure 1-13 illustrates a block diagram for a monochrome computer monitor. Notice that three pieces of information are needed: the video signal, the horizontal sync, and the vertical sync. Since this particular monitor is a monochrome design, only one TTL video signal is needed to operate the *video drive circuit,* after all, pixels are only going to be on or off. The video signal is amplified by a simple transistor switch which drives the CRT's video control grid. There is not a great deal of sophistication here. A contrast control affects the separation between light and darkness by adjusting the amount of amplification given to the video signal.

The *vertical drive circuit* controls the up/down position of the electron beam. A vertical sweep oscillator (54 to 72 Hz depending on the monitor type) provides a ramping signal, which resembles a sawtooth wave. The vertical ramp is triggered by the presence of a Vsync pulse. This vertical signal is amplified by an output driver transistor circuit which connects directly to the vertical deflection

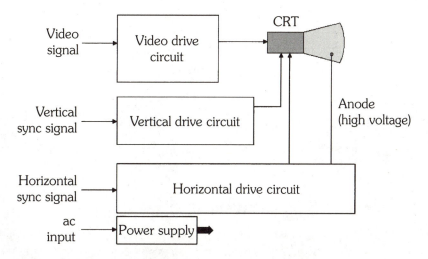

■ **1-13** *Simplified block diagram of a monochrome monitor.*

coil (or *yoke*) fixed to the CRT's neck. The magnetic field produced by the vertical ramp signal maneuvers the electron beam. As the ramp starts from zero and increases in amplitude, the beam moves down the screen.

The *horizontal drive circuit* controls the left/right position of the electron beam. A horizontal oscillator (15.85 to 38 kHz or higher depending on the monitor type) provides a short square wave pulse. The horizontal pulse is triggered by the presence of an Hsync pulse. This horizontal pulse signal is amplified by an output driver transistor circuit which connects directly to the horizontal deflection yoke fixed to the CRT's neck. As with the vertical system, the magnetic field produced by the rapid horizontal signal maneuvers the electron beam. You may wonder why a square wave is used rather than a ramp signal. In truth, horizontal sweeps are so fast that a ramp signal is pointless; component characteristics in the horizontal yoke circuit cause the beam to move in a linear fashion. Keep in mind that there will be many more horizontal than vertical sweeps. For a VGA monitor running at 640 × 480 resolution, there will be 480 horizontal sweeps for every vertical sweep.

Notice that the horizontal drive circuit also controls the monitor's high-voltage system. Remember that thousands of volts are needed to maintain an electron stream. That high voltage is produced by a device known as a *flyback transformer* (FBT). The FBT "spikes" a relatively low voltage up to a much higher level; the principle is similar to the ignition system used in a car. A high-voltage pulse is developed during the horizontal retrace (the brief time when the beam has finished drawing one line and is turned off to be repositioned for the next line). The flyback transformer's primary winding is coupled to the horizontal output. As the horizontal ramp increases, the FBT charges. When the horizontal ramp drops to zero, it does so almost instantly, and this drives the transformer's output to a very high level. Flyback voltages can run anywhere from 15,000 to 30,000 V dc. Later chapters will describe power supplies and flyback operation in much more detail. You will find that many of the problems encountered with modern monitors are related to trouble with the horizontal output driver and flyback circuit.

Inside today's liquid crystal displays

Flat-panel displays have become reliable and efficient display devices for energy-saving desktop systems as well as virtually every laptop, notebook, subnotebook, and pen computer. Liquid crystal displays have certainly earned a place in any discussion of computer monitors. The block diagram for a passive matrix LCD system is illustrated in Fig. 1-14. The heart of the system is the LCD

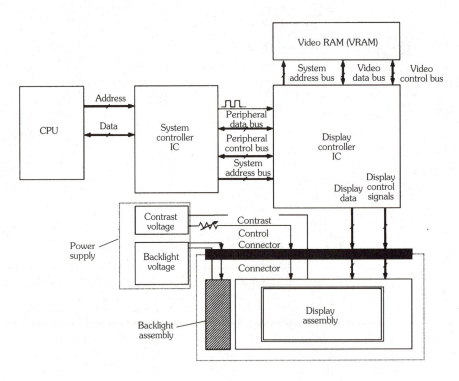

■ **1-14** *Block diagram of a flat-panel display system.*

matrix itself. This is little more than a layer of synthetic gel sandwiched between two glass plates. The glass is etched with an array of horizontal and vertical lines. When the horizontal and vertical drivers are activated, the etched lines connected to the drivers become energized. The point at which the horizontal and vertical lines cross becomes opaque, and this is an LCD pixel.

The LCD matrix is operated by a highly integrated LCD controller integrated circuit (IC) that reads the image data from video memory and generates the "raster" by sequentially firing each horizontal driver and then activating the vertical drivers for each active pixel in the line. This way, the LCD is updated one horizontal line at a time. The LCD controller also communicates with other control elements of the laptop or notebook computer. As you might imagine, LCD assemblies are far simpler than CRT monitors. There are no video or sync signals to contend with because the raw data are being taken from video RAM directly in digital form. There are also no bulky flyback circuits or oscillators to worry about. The only high voltage that is needed is several hundred volts to power the LCD's backlight assembly. Chapter 9 will present a complete discussion of LCD systems and troubleshooting.

21

2

CRT monitor basics

THE CRT COMPUTER MONITOR IS REALLY A MARVEL OF engineering (Fig. 2-1). Although the components have improved dramatically in the last few decades, the monitor works using principles of television receivers that have remained virtually unchanged since the 1950s. There are very few devices in use today that so clearly demonstrate the relationship between physics and electronics. A monitor's job is relatively simple: to produce a visual representation of the computer system's video memory. The monitor accomplishes this by sweeping a regular horizontal and vertical scanning pattern (called *raster*) across the CRT's face while feeding TTL or analog image (pixel) data to the CRT's electron gun(s). Synchronization pulses provided with the image data ensure that each pixel is precisely positioned. This chapter explains the operating principles behind monochrome and color monitors in detail, shows you how they are assembled, and describes the hazards and human factors involved.

If you are already familiar with monitor theory, feel free to skip ahead in the book. You should also keep in mind that the various circuit fragments that are presented here (and throughout the book) are intended as examples only, because it is impossible to represent every possible circuit variation or manufacturer-specific feature. The circuit designs and component values used in your own monitor will likely differ, but the ultimate functions performed by each circuit will be very similar.

Monochrome monitors

The word *monochrome* technically means "one color." Monochrome monitors are capable of displaying images in a single color as well as black (the absence of that color). Since only one color can be produced, the graphic capabilities are very limited. It is possible to *dither* or crosshatch areas of an image to simulate several different shades, but that is a task performed by the video adapter, not the monitor. The most popular use for monochrome

■ **2-1** *An NEC MultiSync 4FGe monitor. (NEC Technologies, Inc.)*

monitors has been as basic ASCII text displays. Text-intensive applications such as library reference, airline bookings, and on-line information services still continue to use monochrome displays with great success; they certainly are not obsolete.

In the last chapter, you saw a monochrome monitor depicted as a basic block diagram in Fig. 1-13. This chapter will expand on each of those simple blocks in Fig. 2-2, and explain the detailed operation of each section. Generally speaking, the monitor's operation can be broken down into five major operating areas: the CRT, the video drive circuit, the vertical drive circuit, the horizontal drive/high-voltage circuit, and the power supply.

The monochrome CRT

The *cathode-ray tube* (CRT) is basically a large vacuum tube which is filled with an inert gas under very low pressure (Fig. 2-3). The presence of inert gas and low pressure allow electrons to flow with no interference. The *cathode* (a negative electrode) is energized by a 6.3-V signal at 450 mA (some CRTs use up to 15 V or more to energize the cathode). This causes electrons to "boil off" the cathode. Simultaneously, a high-voltage positive potential is applied to the CRT's large rectangular face which acts as the *anode* (a positive electrode). The difference in potential

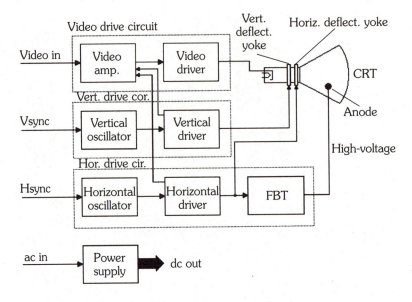

■ **2-2** *Block diagram of a monochrome monitor.*

between the cathode and anode causes electrons liberated by the cathode to stream toward the anode. When electrons strike the front face (the screen), the kinetic energy of the electrons causes phosphor to illuminate at the point of impact. The color of the illumination is determined by the phosphor coating being used (the electrons themselves are invisible). By changing the

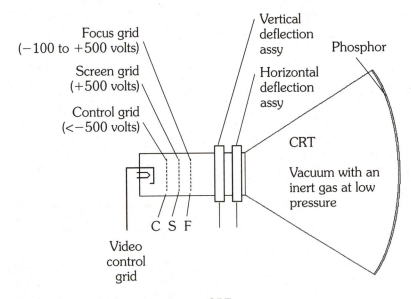

■ **2-3** *Diagram of a monochrome CRT.*

phosphor chemistry, CRTs with different colors can be manufactured. An interesting property of phosphors is their *persistence,* that is, how long the phosphor point will glow after being struck by the electron beam. Table 2-1 lists a series of phosphor types, colors, and persistence times used by typical monochrome CRTs. You will notice that many colors fall into three categories: green (the traditional CRT color), amber (rich yellow, a de facto European standard), and white.

You should also be aware of the *CIE Chromaticity Chart* (also known as the *Kelly Chart of Color Designation for Lights*). While the phosphor chart in Table 2-1 lists specific colors, it is very difficult to discern the particular shade. The CIE Chromaticity Chart illustrates a range of colors throughout the visible spectrum. Unfortunately, we could only reproduce a black and white representation of the chart as shown in Fig. 2-4, but actual full-color

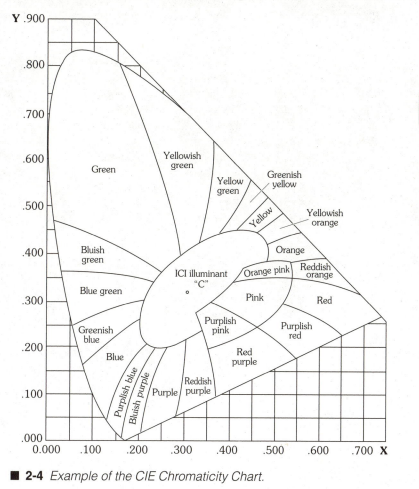

■ **2-4** *Example of the CIE Chromaticity Chart.*

■ Table 2-1 Typical phosphor characteristics for video displays

New type	Old type	Color	CIE (X/Y)	Persistence (ms)
VA	P1	Yellow-green	.221/.713	n/a
AM	P4	White	.268/.294	0.100
7M	P11	Blue	.144/.143	n/a
3A	P19	Orange	.569/.427	n/a
MY	P31	Yellow-green	.266/.538	0.070
3C	P38	Orange	.562/.436	n/a
VM	P39	Yellow-green	.201/.716	0.070
AN	P40	White	.259/.322	0.045
MM	P42	Yellow-green	.253/.558	0.100
MN	P43	Yellow-green	.327/.551	1.500
DA	P45	White	.256/.311	n/a
MA	P46	Yellow-green	.396/.564	n/a
7N	P55	Blue	.148/.063	n/a
7Y	P56	Red-orange	.639/.352	n/a
RM	PC101	Yellow-green	.210/.685	0.125
—	P103	White	n/a	0.084
AP	PC104	White	.280/.304	0.085
1M	P106	Orange	.571/.400	0.300
—	P108	Yellow-green	n/a	0.125
—	P109	Yellow-green	n/a	0.080
MP	PC110	Yellow-green	.253/.586	0.080
AQ	PC115	White	.319/.349	n/a
7A	PC116	Yellow-green	.201/.716	n/a
AR	PC118	White	.259/.260	0.090
VB	PC122	Yellow-green	.211/.715	0.075
3M	P134	Orange	.555/.444	0.500
—	P136	White	n/a	0.085
RN	PC137	Yellow-green	.210/.700	0.125
MQ	PC138	Yellow-green	.261/.540	0.070
MZ	PC139	Yellow-green	.280/.591	0.070
MC	PC141	Yellow-green	.253/.551	0.100
—	P143	White	n/a	0.050
—	P146	Yellow-green	n/a	0.080
YA	PC148	Yellow-green	.207/.711	n/a
—	P154	Yellow-green	n/a	0.075
YY	PC156	Yellow-green	.280/.553	n/a
VD	PC158	Green-yellow	.462/516	n/a
YB	PC159	Yellow-green	.214/.709	n/a
M2	PC160	Yellow-green	.271/.564	0.070
—	P161	Yellow-green	n/a	0.070
MR	PC162	Yellow-green	.222/.661	0.100
KA	PC163	White	.329/.350	n/a
GA	PC164	White	.254/.282	0.100

New type	Old type	Color	CIE (X/Y)	Persistence (ms)
AS	PC167	White	.343/.381	0.075
M3	PC168	Green-yellow	.282/.617	n/a
MS	PC169	Yellow	.457/.474	1.500
1A	PC170	Orange	.571/.400	n/a
GB	PC171	White	.357/.391	0.200
72	PC175	Red-orange	.639/.351	0.600
M4	PC178	Yellow-green	.270/.537	0.100
—	P179	White	n/a	1.000
—	P180	Yellow-orange	n/a	0.075
YM	PC181	Yellow-green	.406/.503	n/a
—	P182	Orange	n/a	0.500
AT	PC184	White	.316/.341	0.075
—	P185	Orange	n/a	0.300
VN	PC186	Yellow-green	.205/.717	0.250
DM	PC188	White	.352/.382	0.050
—	P190	Orange	n/a	0.100
GC	PC191	White	.355/.395	0.120
GD	PC192	White	.334/.377	0.200
AY	PC193	White	.315/.350	0.080
1N	PC194	Orange	.549/.440	0.170
AU	PC195	White	.348/.387	n/a
6A	PC197	Orange-yellow	.545/.444	n/a
1B	PC198	Orange	.549/.440	n/a
1C	PC199	Orange	.580/.415	n/a
7B	PC200	Yellow	.499/.482	n/a
YN	PC202	Yellow-green	.229/.708	n/a
ME	PC203	Yellow-green	.452/.480	n/a
RB	PC204	Green-yellow	.451/.510	n/a
7C	PC205	Yellow	.500/.481	n/a
AV	PC206	White	.352/.373	n/a
GE	PC208	White	.363/.394	n/a
7D	PC209	Yellow	.446/.518	n/a
MT	PC210	Yellow	.478/.478	n/a
DN	PC211	White	.323/.357	n/a
GF	PC212	White	.340/.367	n/a
MF	PC213	Green-yellow	.460/.510	n/a
MU	PC214	Yellow-green	.199/.717	n/a
GG	PC215	White	.356/.378	n/a
GH	PC216	White	.255/.278	n/a
MV	PC217	Yellow-green	.268/.532	n/a
1Y	PC218	Orange	.557/.437	n/a
GK	PC219	White	.348/.386	n/a

charts can be found in engineering and design texts. You can find any particular color using an X/Y reference scheme. For example, consider the ac (PC143) phosphor shown in Table 2-1. The CIE coordinates are listed as .262 by .296. If you find the intersection of those coordinates on the chart, the color at that location gives you a pretty good idea of actual phosphor color.

Another phosphor consideration relates to *aging* through regular use. Even when run within the proper limits, screen phosphors will not last forever. Over time, phosphors will darken, and their light output (efficiency) will decline. This symptom is most noticeable as phosphor "burn-in," where an image displayed for prolonged periods will eventually degrade and burn the phosphor to the point where you can actually see discolored areas on the CRT when the monitor is turned off. As a technician, there is absolutely nothing you can do to correct phosphor aging except replace the CRT outright. The most effective means of combating phosphor aging is to take protective measures by keeping screen brightness down and blanking the display when the monitor is not in use (e.g., turn the monitor off or use a screen saver).

Of course, the flow of electrons must be controlled to produce a meaningful display. Otherwise, a CRT would just shoot a dim column of electrons that would appear as a dull, unfocused cloud in the center of the CRT. The beam must be focused, regulated, and directed around the screen as required. Beam intensity is modulated by the *video control grid*. A grid is just what the name implies: a fine mesh of wire inserted in front of the cathode. As the amplified video signal reaches the video control grid, the strength of the electron beam passing the grid can be changed. For a monochrome CRT, the video signal is either on or off, so the beam is either virtually cut off or on at full power.

Once the beam intensity is modulated by the video control grid, a series of three main grids maintains the beam quality. First, a *control grid* serves as a "master valve" to regulate the flow of electrons. As you see in Fig. 2-5, the control grid affects the screen brightness which can be optimized through the use of a brightness control. A *screen grid* provides several hundred volts of positive potential which begin to accelerate the electrons. As the beam picks up speed and passes the screen grid, it is still a rather loose column of electrons. The beam must be focused before it is accelerated to the screen. Focus is accomplished by a *focus grid,* which provides a −119- to +490-V potential that acts to condense the electrons into a narrow line.

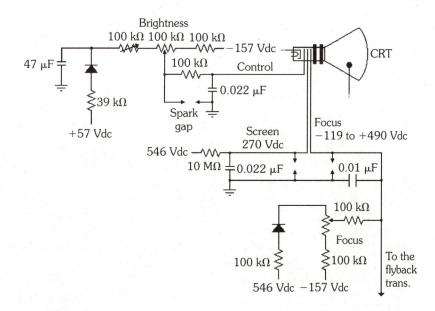

■ **2-5** *CRT grid control circuit fragment.*

At this point, the narrowly focused beam must be directed around the screen. A set of powerful magnetic fields is used to orchestrate the beam path. One magnetic field is applied in the vertical direction to move the beam up and down, and another magnetic field is applied in the horizontal direction to move the beam left and right. Magnetic fields are produced by *deflection yokes* which are little more than coils of fine wire placed precisely over the CRT's neck. Signals from the monitor's vertical and horizontal drive circuits produce the *raster* (or scanning pattern) that continuously deflects the beam. Remember that there are no moving parts in the CRT.

Video drive circuit

Ultimately, the task of a video drive circuit is to regulate the strength of an electron beam. A video drive circuit must convert a small video signal into a signal large enough to drive the CRT's video control grid. For monochrome monitors with a single TTL video line, the video drive circuit turns the electron beam either on or off. For monochrome monitors with two TTL inputs, one line turns the beam on or off, while the second line sets the beam intensity to a high or low level. A second input to control intensity adds a small amount of diversity to the display.

Figure 2-2 illustrates the block diagram for a monochrome video stage. An initial amplifier stage translates small on/off signals into true TTL logic levels, which are further conditioned by conventional logic circuitry. A *contrast control* is typically added to the initial amplifier stage which allows a user to adjust the difference in on/off intensities. Unfortunately, the amplified TTL signal is not enough to drive the CRT's video control grid. Additional amplification is required, so a secondary amplifier stage is added. A power amplifier circuit boosts the TTL signal to a level of 20 to 40 V.

Problems can strike the video drive circuit in a number of ways, but there are clues to help guide your way. If the display should disappear, but the raster remains (raster is that dim haze you see by turning up the monitor's brightness), the video signal may have failed at the video adapter board or the monitor's video amplifier circuit may have quit. Try a monitor known to be in good condition. If the correct image appears, you know the video adapter is producing the desired output, and the original monitor is probably defective. If no display appears on the good monitor, suspect the video adapter board in your PC. If the screen is black, suffers from fixed brightness (with or without video input), or loses focus, one or more grids in the CRT may have shorted and failed. Refer to the chapters on monitor troubleshooting for more detailed instructions.

Vertical drive circuit

The main task of the vertical drive circuit is to operate the vertical deflection yoke. This is accomplished with a *vertical sweep oscillator,* which is little more than a free-running oscillator set to run at 50 to 60 Hz. The actual oscillator may be based on either a transistor or an integrated circuit depending on the monitor's age. When a vertical synchronization trigger pulse is received from the video adapter board, the oscillator is forced to fire. When the oscillator is triggered, it produces a sawtooth wave similar to the one shown in Fig. 2-6. The start of the sawtooth wave corresponds to the top of the screen, while the end of the sawtooth wave corresponds to the bottom of the screen. When the sawtooth cycle is complete, there is a blank period for blanking and retrace. One vertical sweep will be accomplished in less than 1/50th of a second.

However, the sawtooth signal generated by the vertical oscillator circuit does not have enough power to drive the vertical deflection yoke directly. The vertical deflection sawtooth signal is amplified by a *vertical output driver* circuit that provides significant current needed to induce strong magnetic fields in the vertical deflection yoke. The amount of power needed dictates the use of a high-power transistor arrangement.

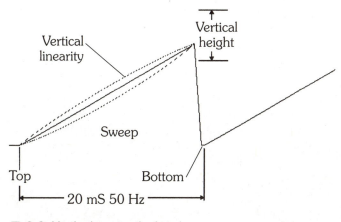

■ 2-6 *Vertical sawtooth signal.*

There are a number of adjustments in the vertical drive circuit that you should be familiar with. The *vertical linearity* optimizes the shape of the sawtooth waveform. Ideally, the wave's upward ramp should be perfectly straight (or linear). In actual practice, however, the slope of the line may vary a bit from start to finish. This translates to the display. Since the ramp defines the spacing between individual lines, any variation in the ramp "slope" will affect the spacing between horizontal lines. Some lines may be too far apart, whereas others may appear too close together. Typically, you need not adjust vertical linearity unless you replace a component in the oscillator circuit. The *vertical size* control adjusts the slope of the ramp signal, which affects the final amplitude of the sawtooth. This effectively compresses or expands the screen image in the vertical orientation.

Trouble with the vertical drive circuit usually strikes the vertical output driver circuit. The output drive is often designed as a "push-pull" amplifier. If either half of the push-pull configuration should fail, the corresponding half of the image will disappear. If both halves of the push-pull configuration should fail, the screen image will compress to a straight horizontal line in the center of the screen (there would be no vertical deflection, only horizontal deflection). Another problem is vertical *oversweep* which elongates the picture to the extent that it "wraps back" on itself in the lower portion of the screen. The area where the vertical image oversweeps will appear with a whitish haze and is typically the fault of the vertical oscillator circuit.

Horizontal drive circuit

The main task of the horizontal drive circuit is to operate the horizontal deflection yoke. This is accomplished with a *horizontal sweep oscillator,* which is little more than a free-running oscillator set to run at a frequency between 15 and 40 kHz (or higher depending on the monitor's age). A monochrome monitor will typically use a horizontal sweep frequency of about 15.75 kHz. The actual oscillator may be based on a transistor, but it is usually designed around an integrated circuit, which is more stable at the higher frequencies that are needed. When a horizontal synchronization trigger pulse is received from the video adapter board, the oscillator is forced to fire. When the oscillator is triggered, it produces a square wave. The start of the square wave corresponds to the left side of the screen. When the horizontal cycle is complete, there is a blank period for blanking and retrace. One horizontal sweep will be accomplished in less than 63.3 µs.

As with the vertical oscillator, however, the pulse signal generated by the horizontal oscillator circuit does not have enough power to drive the horizontal deflection yoke directly. The horizontal deflection signal is amplified by a *horizontal output driver* circuit which provides significant current needed to induce strong magnetic fields in the horizontal deflection yoke. The amount of power needed dictates the use of a high-power transistor arrangement.

There are a number of adjustments in the horizontal drive circuit that you should be familiar with. The *horizontal linearity* optimizes the characteristics of the horizontal signal. Ideally, the rapid left-to-right sweep of the electron beam should be the same speed from start to finish. This rate defines the spacing between individual pixels, so any variation in the sweep rate will affect the spacing between pixels. In some areas, pixels may be too far apart, whereas pixels in other areas may appear too close together. Typically, you need not adjust horizontal linearity unless you replace a component in the horizontal oscillator circuit. The *horizontal size* control adjusts the sweep duration. This effectively compresses or expands the screen image in the left-to-right orientation. A third control allows the adjustment of *horizontal centering* by introducing a slight delay between the time a horizontal synchronization (Hsync) pulse is received and the time a sweep pulse is started. By default, there is always some delay needed to produce a centered display image. Reducing the delay moves the screen image to the left, while increasing the delay moves the screen image to the right. Centering and size controls are most useful for optimizing the image size and position for a particular video mode (typical text or low-resolution graphics for monochrome systems) and should not need readjustment unless the video mode changes.

Trouble with the horizontal drive circuit usually strikes the horizontal output driver circuit because that is the circuit that sustains the greatest stress in the monitor. The horizontal output drive is usually designed as a dual-transistor switching circuit using two power transistors. If either of the power transistors should fail, the entire image will disappear since high voltage will also be affected. Unfortunately, a fault in the horizontal sweep oscillator will also result in an image loss because high-voltage generation depends on a satisfactory horizontal sweep pulse. If the horizontal oscillator or amplifiers fail, high voltage fails as well, and the image becomes too faint to see. This makes troubleshooting horizontal problems a bit more difficult than troubleshooting vertical problems.

High-voltage circuit

Technically, the high-voltage system is part of the horizontal drive circuit, but we discuss it separately because of its particular importance in computer monitors. The monitor's power supply generates relatively low voltages (usually not much higher than 140 V), so the high positive potential needed to excite the CRT's anode is not developed in the power supply. Instead, the 15 to 30 kV needed to operate a CRT is generated from the horizontal output. The amplified high-frequency pulse generated by the horizontal output driver circuit is provided to the primary winding of a device known as the *flyback transformer* (FBT), which is the heart of a monitor's high-voltage system.

Figure 2-7 illustrates the importance of the flyback transformer. There are three secondary windings on an FBT. The lower winding is a simple step-down winding that provides a low ac voltage (typically about 6.2 to 15 V ac depending on the particular CRT) which heats the CRT's cathode. The middle winding provides about 150 V ac to the CRT control circuit. A 500-V signal from the horizontal output and a 53-V input from the monitor's power supply also power the CRT control circuit. The circuit itself generates the three main CRT control grid voltages that you saw in Fig. 2-3: the adjustable control grid voltage (brightness), the adjustable focus grid voltage (focus), and the fixed screen grid voltage.

The top winding is perhaps the most vital—it is the high-voltage winding that steps up the horizontal signal to the required 15- to 30-kV level. Notice that a high-voltage diode is placed in series with the high-voltage winding to *rectify* the high ac level to a dc level. The 500 pF of effective capacitance found in the CRT assembly acts to *filter* (or smooth) the high voltage into a useful

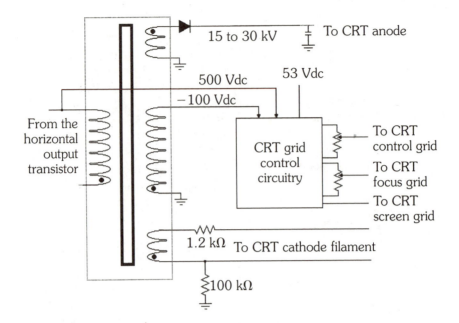

15 to 30 kV To CRT anode

500 Vdc 53 Vdc

−100 Vdc

From the horizontal output transistor

CRT grid control circuitry

To CRT control grid

To CRT focus grid

To CRT screen grid

1.2 kΩ To CRT cathode filament

100 kΩ

■ **2-7** *Simplified diagram of a flyback circuit.*

form. As a result, the monitor's high-voltage circuit forms a crude power supply of its own. You will learn much more about rectification and filtering in Chapter 6.

It is not difficult to see that a fault in the horizontal output drive circuit or the flyback transformer could disable the entire monitor. It is very common for trouble to strike the flyback transformer, especially the high-voltage windings. As transformers age, the protective insulation breaks down, and high voltage can arc between windings. In its initial stages, this may result in a loss of picture brightness. For severe breakdowns, however, the image may disappear entirely. Such breakdowns are also accompanied by a high-pitched squealing or hissing sound indicating a high-voltage electrical arc. *Defective components in the high-voltage system **must** be replaced with exact component values.*

Power supply

Of course, the foundation of every monitor's operation is the power supply itself. Commercial ac is converted into a series of relatively low dc voltages that power the monitor. A monochrome monitor will typically employ a power supply that delivers +5, +15, and +60 V dc, but be aware that there will be variations in the supply's outputs depending on the design of each particular monitor. For example, a typical color monitor may use a power supply that delivers +6.3, +12, +20, +87, and +135 V dc. It is advisable to inspect the outputs from your particular power supply circuit to determine the exact number of outputs and each output level. Service data for your particular monitor will also specify the power supply output levels and their locations.

Color monitors

Once the PC gained acceptance in the marketplace, customer demand spurred the development of color displays (a demand that continues to this day). The reasons for color are simple. First, color is easier on the eyes, especially when you have to watch a monitor for prolonged periods. Color images are attractive, so every application from games to spreadsheets can use color to develop more pleasant and intuitive screens. Finally, color displays open up new applications such as image processing and multimedia, which are simply impossible with monochrome monitors.

The step up from monochrome to color comes at a price, however. Much of the raster circuitry in a color monitor is identical to that of a monochrome monitor, but color video processing circuitry is more extensive and sophisticated as you see in the block diagram of Fig. 2-8. Three separate video drive circuits are needed to handle the red, green, and blue primary color signals. By combining the primary color signals at the CRT, virtually any color can be produced. This part of the book explains how color monitors work and how color operation relates to monochrome monitors.

The color CRT

Although color monitors rely on extra video circuitry to process color signals, it is the design and construction of the CRT itself that really make color monitors possible. The basic principles of a color CRT (Fig. 2-9) are very similar to a monochrome monitor—electrons "boil off" the cathode and are accelerated toward the phosphor-coated front screen by a high positive potential. Color CRTs use *three* cathodes and video control grids with one for each of the three primary colors. Control (brightness), screen, and focus grids serve the same purpose as they do in monochrome CRTs. The *control grid* regulates the overall brightness of the electron beams, the *screen grid* begins accelerating the electron beams toward the front screen, and the *focus grid* narrows the beams. Once the electron beams are focused, vertical and horizontal deflection coils (or deflection yokes) apply magnetic force to direct the beams around the screen.

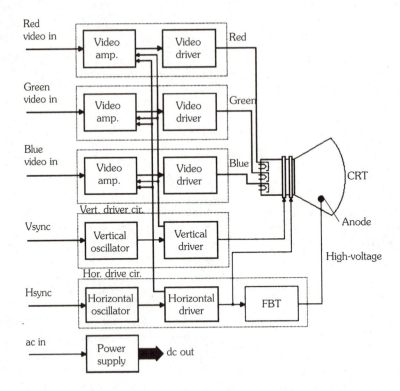

■ 2-8 *Block diagram of a color monitor.*

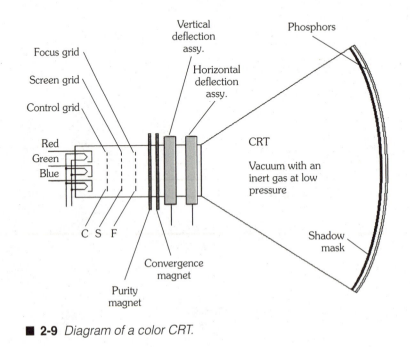

■ 2-9 *Diagram of a color CRT.*

You will notice a *shadow mask* added to the color CRT. A shadow mask is a thin plate of metal that contains thousands of microscopic perforations with one perforation for each screen pixel. The mask is placed in very close proximity to the phosphor face. There is also a substantial difference in the screen phosphors. Where a monochrome CRT uses a homogeneous layer of phosphor across the entire face, a color CRT uses phosphor *triads* as shown in Fig. 2-10 (the distance between the shadow mask and phosphor screen is shown greatly exaggerated). Red, green, and blue phosphor dots are arranged in sets such that the red, green, and blue electron beams will strike the corresponding phosphor. Of course, the electron beams are invisible (and color is determined by the phosphor itself), but each electron beam is intended to excite only one color phosphor. In actual operation, the color dots are so close together that each triad appears as a single point (or pixel).

Color CRTs must also be more precise in how the three electron beams are directed around the screen. Since there are now three phosphors instead of just one, it is critical that each electron beam strike only its corresponding phosphor color, not adjoining phosphors. This is known as *color purity*. A *purity magnet* added to the CRT yoke helps to adjust fine beam positioning. By using a shadow mask, the electron beams are only allowed to reach the phosphors where there are holes in the mask. Also realize that each of the three electron beams must converge at each hole in the shadow mask. A *convergence magnet* added to the CRT yoke adjusts beam convergence in the display center (known as *static convergence*), and a convergence coil driven by the raster circuitry optimizes beam convergence at the edges of the display

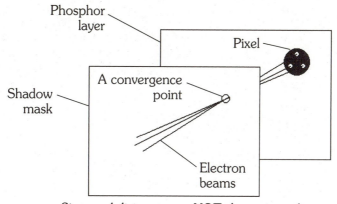

Sizes and distances are NOT shown to scale

■ **2-10** *Use of a shadow mask in a color CRT.*

(known as *dynamic convergence*). It is this delicate balance of purity and convergence adjustments, as well as the presence of a shadow mask, that gives today's color monitors such rich, precise color.

Video drive circuits

A video drive circuit is used to regulate the strength of the electron beam by adjusting the signal strength on the corresponding video control grid in the CRT. The video drive circuit must convert a small video signal (usually no more than 0.7 V peak to peak) into a signal large enough to drive the CRT (typically around 50 V). For color monitors with three analog video lines, three separate video drive circuits are required. Where a monochrome monitor simply turns the electron beam either on or off, a color monitor must handle faint, rapidly changing analog signals.

In principle, each video stage is remarkably similar to its monochrome counterpart. In actual practice, however, color video stages are somewhat more complex. The initial video amplifier of each stage (often referred to as a *preamp*) is a linear, high-gain amplifier which translates the low-level analog color signal into a clear, solid analog level of several volts (usually about 3 V). A common *contrast control* is typically added which affects each of the three video amplifiers simultaneously. Unfortunately, the amplified analog signal is not enough to drive the CRT's video control grid. Additional amplification is required, so a secondary amplifier stage is added. Some monitor designs use a three-stage amplifier circuit for each color: a high-gain *preamp* for lots of amplification at little power, a second *amplifier* that provides less amplification but more power to the signal, and a *driver* that offers almost no amplification but provides enough signal power to drive the CRT.

Problems can strike the video drive circuits in a number of ways, but there are clues to help guide your way. If the display should disappear, but the raster remains (raster is that dim haze you see by turning up the monitor's brightness), the video signal may have failed at the video adapter board in your PC. If there is suddenly not enough (or far too much) red, green, or blue in the displayed image, the corresponding digital-to-analog converter (DAC) on the video adapter may have failed or the corresponding video drive circuit in the monitor may have broken down. Try a monitor known to be in good condition. If the correct image appears, you know the video adapter is producing the desired output, and the original monitor is probably defective. If no display appears on the good monitor, suspect the video adapter board in your PC. If the screen is black, suffers from fixed brightness (with or without video input), or loses focus, one or more grids in the CRT may have shorted and failed. Refer to the chapters on monitor troubleshooting for more detailed instructions.

Vertical drive circuit

The vertical drive circuit is used to operate the vertical deflection yoke and is part of the color monitor's overall raster circuitry. This is accomplished with a *vertical sweep oscillator*, which is little more than a free-running oscillator set to run at either 60, 70, 72 Hz (or more depending on the design of the particular monitor). Older color monitors typically used transistor-based oscillators, but virtually all current color monitors use an integrated circuit (IC) oscillator because ICs provide a very linear and precise signal. As with monochrome monitors, the oscillator is forced to fire when a vertical synchronization trigger pulse is received from the video adapter board. When the oscillator is triggered, it produces a sawtooth wave similar to the one shown in Fig. 2-11. The start of the sawtooth wave corresponds to the top of the screen, while the end of the sawtooth wave corresponds to the bottom of the screen. When the sawtooth cycle is complete, there is a blank period for blanking and retrace. One vertical sweep will be accomplished in less than 1/60th of a second (or 1/70th or 1/72nd of a second depending on the monitor).

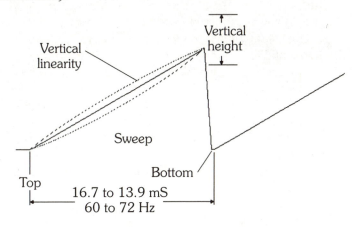

■ **2-11** *A vertical sawtooth signal in a color monitor.*

However, the sawtooth signal generated by the vertical oscillator circuit does not have enough power to drive the vertical deflection yoke directly. The vertical deflection sawtooth signal is amplified by a *vertical output driver* circuit. The vertical amplifier provides the significant current needed to induce a strong magnetic field in the vertical deflection yoke. The amount of power needed to operate the deflection yoke demands the use of a high-power transistor arrangement, but many current monitor designs are using high-power amplifier ICs instead of discrete transistors.

There are a number of adjustments in the vertical drive circuit that you should be familiar with. The *vertical linearity* optimizes the shape of the sawtooth waveform. Ideally, the wave's upward ramp should be perfectly straight (or linear). In actual practice, however, the slope of the line may vary a bit from start to finish. This translates to the display. Since the ramp defines the spacing between individual lines, any variation in the ramp "slope" will affect the spacing between horizontal lines. Typically, you need not adjust vertical linearity unless you replace a component in the oscillator circuit. The *vertical size* control adjusts the slope of the ramp signal, which affects the final amplitude of the sawtooth. This effectively compresses or expands the screen image in the vertical orientation. By applying a dc offset to the vertical deflection signal, the raster can be centered in the display with a *vertical centering* control.

Trouble with the vertical drive circuit usually strikes the vertical output driver circuit. The output drive is usually designed as a "push-pull" amplifier (even IC-based power amplifiers use such a configuration internally). If either half of the push-pull configuration should fail, the corresponding half of the image will disappear. If both halves of the push-pull configuration should fail (or the oscillator should fail), the screen image will compress to a straight horizontal line in the center of the screen (there would be no vertical deflection, only horizontal deflection). The best way to deal with such a problem is to test and replace the amplifier transistors or IC. Another problem is vertical oversweep which elongates the picture to the extent that it "wraps back" on itself in the lower portion of the screen. The area where the vertical image oversweeps will appear with a whitish haze and is typically the fault of the vertical oscillator circuit. Refer to the chapters on monitor troubleshooting for more detailed information.

Horizontal drive circuit

The horizontal drive circuit is the second part of the color monitor's raster circuit, and it is designed to operate the horizontal deflection yoke. This is accomplished with a *horizontal oscillator* which is little more than a free-running oscillator set to run at a frequency between 15 and 48 kHz (or higher). A CGA monitor will typically use a horizontal sweep frequency of about 15.75 kHz. The actual oscillator may be based on a transistor, but it is usually designed around an integrated circuit, which is more stable at the higher frequencies that are needed. When a horizontal synchronization trigger pulse is received from the video adapter board, the oscillator is forced to fire. When the oscillator is triggered, it produces a square wave. The start of the square wave corresponds to the left side of the screen. When the cycle is complete, there is a blank period for blanking and retrace. At an operating frequency of 31.5 kHz, one horizontal sweep will be accomplished in about 31.7 µs.

As with vertical oscillator circuits, the signals generated by the horizontal oscillator do not have enough power to drive the horizontal deflection yoke directly. The horizontal deflection pulse signal is amplified by a *horizontal output driver* circuit which provides the significant current needed to induce strong magnetic fields in the horizontal deflection yoke. The amount of power needed dictates the use of a high-power transistor arrangement. Transistors are still popular today as horizontal output drivers.

There are a number of adjustments in the horizontal drive circuit that you should be familiar with. The *horizontal linearity* optimizes the shape of the horizontal sweep. Ideally, the left-to-right sweep rate should be perfectly even, but the actual sweep rate may vary a bit from start to finish. This variation will affect the spacing between pixels. Typically, you need not adjust horizontal linearity unless you replace a component in the oscillator circuit. The *horizontal size* control adjusts the signal magnitude, which affects the amount of sweep applied to the electron beam. This effectively compresses or expands the screen image in the left-to-right orientation. A third control allows the adjustment of *horizontal centering* by introducing a slight delay between the time a horizontal synchronization (Hsync) pulse is received and the time a horizontal pulse is generated. By default, there is always some delay needed to produce a centered display image. Reducing the delay moves the screen image to the left, while increasing the delay moves the screen image to the right. Centering and size controls are most useful for optimizing the image size and position for a particular video mode and should not need readjustment unless the video mode changes.

Trouble with the horizontal drive circuit usually strikes the horizontal output drive circuit because that is the circuit that sustains the greatest stress in the monitor. The horizontal output drive is usually designed as a dual-transistor switching circuit using two

41

high-power transistors. If the power transistors should fail, the entire image will disappear since high voltage will also be affected. Unfortunately, a fault in the horizontal oscillator will also result in an image loss because high-voltage generation depends on a satisfactory horizontal pulse. If the horizontal oscillator or amplifier fails, high voltage fails as well, and the image becomes too faint to see. This makes troubleshooting horizontal problems a bit more difficult than troubleshooting vertical problems. Refer to the chapters on monitor troubleshooting for detailed information on how to approach horizontal circuit troubleshooting.

High-voltage circuit

As you learned from the discussion of monochrome monitors, the high-voltage system is actually part of the horizontal drive circuit. We discuss it separately in this chapter because of its importance in computer monitors. A monitor's power supply generates relatively low voltages (usually not much higher than 140 V). This means that the high positive potential needed to excite the CRT's anode is **not** developed in the power supply. Instead, the 15 to 30 kV needed to power a CRT anode is generated from the horizontal output. The amplified high-frequency pulse signal generated by the horizontal output driver circuit is provided to the primary winding of a device known as the *flyback transformer* (FBT). The FBT produces the high voltage.

You can see the diagram for a typical FBT in Fig. 2-7; the principles here are the same as a monochrome monitor. There are three secondary windings on an FBT. The lower winding is a simple step-down winding that provides a low ac voltage (typically about 6.2 to 15 V ac depending on the particular CRT) which heats the color CRT's three cathodes. The middle winding provides about 150 V ac to the CRT control circuit. A 500-V signal from the horizontal output and a 53-V input from the monitor's power supply also power the CRT control circuit (the actual voltage levels will vary slightly from monitor to monitor). Adjustable control grid voltage (brightness), the adjustable focus grid voltage (focus), and the fixed screen grid voltage are all generated by the FBT and circuit similar to Fig. 2-7.

The top winding is the high-voltage winding that steps up the horizontal signal to the required 15- to 30-kV level (the actual voltage level will depend on the particular monitor with larger CRTs requiring greater voltages to operate). Notice that a high-voltage diode is placed in series with the high-voltage winding to *rectify* the high ac level to a dc level. The 500 pF of effective capacitance found in the CRT assembly acts to *filter* (or smooth) the high voltage into a useful form. You will read much more about rectification and filtering for power supplies in Chapter 6.

After looking at Fig. 2-7, it is not difficult to see that a fault in the horizontal output drive circuit or the flyback transformer could disable the entire monitor. It is very common for trouble to strike the flyback transformer, especially the high-voltage windings. As transformers age, the protective insulation breaks down and high voltage can arc between windings. In its initial stages, this may result in a loss of picture brightness. For severe breakdowns, however, the image may disappear entirely. Flyback transformer breakdowns are often indicated by a high-pitched squeal or hiss indicating a high-voltage arc.

Power supply

The operation of every color monitor relies on the proper performance of a power supply where commercial ac is converted into a series of relatively low-dc voltages that power the monitor. A color monitor will typically employ a power supply that delivers +6.3, +12, +20, +87, and +135 V dc, but be aware that there will be variations in the supply's outputs depending on the design of each particular monitor. Chapter 6 discusses power supply operation, troubleshooting, and repair in detail.

Understanding power-saving monitor systems

By themselves, personal computers demand relatively little power, about as much as a family-sized television. As a consequence, many PC users and employers have chosen to leave their PCs turned on all day long (and often overnight). Ten years ago, this would hardly have been an issue, but today there are 10s of millions of PCs in service around the world. In the United States alone, PCs are believed to account for over 5 percent of all commercial electricity consumption! That figure is expected to rise to over 10 percent by the year 2000. Now 10 percent may not seem like much, but the numbers work out to about 22 billion (with a "b") kWh of electricity each year. The cost of that much power provides a strong incentive for conservation, and the PC industry has been working hard to respond.

The video breakthrough

The one notable part of the PC to cut through the delays and confusion of power conservation has been the *video system.* This is particularly important when you realize that the monitor itself usually consumes far more power than the desktop system it is attached to, and the conservation can be accomplished without any other "green" components in the system. The success (and overall ease) of video power conservation is largely due to the fact that monitors and video boards are a relatively remote and independent part of the PC. After all, you can swap a monitor or video board in a matter of minutes with almost no worry of incompatibility, unlike updating a motherboard or power supply. The following sections show you the concepts behind video power conservation and explain how to implement a "green" video system.

Energy star compliance

If you choose to work with energy-saving video equipment, you are invariably going to encounter *Energy Star,* a program sponsored by the U.S. Environmental Protection Agency (EPA). Energy Star is not a *standard* or *specification;* rather it is a series of minimum requirements outlined for PC and peripheral power consumption (this includes monitors as well as printers and other peripheral components). When a device meets the minimum requirements, it is said to be "Energy Star compliant" and has earned the right to bear the EPA's Energy Star logo.

Very simply, Energy Star compliance requires that PCs, monitors, and printers must be able to enter a power-saving mode (also referred to as a *low-power, sleep,* or *standby* mode) when idle for some period of time and then recover from that state when needed without interrupting the machine's state or work left in progress. PCs and monitors must power down to 30 W or less (each). Printers that normally run under 15 pages per minute must power down to 30 W or less and those that run at over 15 pages per minute can power down to 45 W or less. All high-end color printers must power down to 45 W or less.

By all accounts, these requirements are pretty loose. You should realize that the Energy Star program does not dictate "how" a device achieves its energy-saving state, nor does it define any kind of signaling protocol to control energy-saving features. All of that is left up to the individual manufacturers. As a result, all Energy Star actually says is that a device must use less than "such-and-such" an amount of power while in its "power-saving" state. It is also important to remember that the Energy Star program is strictly voluntary within the PC industry; thus, manufacturers are not legally obligated to design or produce energy-saving products.

Conservation techniques

Now that you understand the objectives of video power conservation, it's time to see how the conservation is implemented on a practical basis. While various peripheral manufacturers work to develop their own power-saving schemes, two signaling standards have come to the forefront of video technology: DPMS and Nutek. As power-saving video equipment becomes more readily available, you will probably encounter both techniques.

DPMS

The *Display Power Management System* (DPMS) was originally developed by the Video Electronic Standards Association (known as VESA) to specifically address power conservation in computer monitors. The DPMS protocol uses the horizontal and vertical synchronization signals generated from a video board to set the monitor's power-saving mode.

Ordinarily, horizontal and vertical sync pulses are generated *continuously* from a video board, which produces the *raster* (a lighted area of the screen that the image appears in) that you can see by turning up the screen brightness. The DPMS defines four monitor states: *on, standby, suspend,* and *off.* By selectively cutting off one or both sync signals at the video board, the DPMS monitor can be placed in one of those modes as shown in Table 2-2.

■ **Table 2-2 DPMS state versus video signals**

VESA DPMS	Horizontal sync	Vertical sync	Video
On	normal	normal	normal
Standby	cutoff	normal	blanked
Suspend	normal	cutoff	blanked
Off	cutoff	cutoff	blanked

Unfortunately, DPMS compatibility does not define the power levels at each state. *Therefore, there is no guarantee that a DPMS-compliant monitor is actually Energy Star compliant.* The VESA specifically *avoided* defining power levels with DPMS signaling so that the DPMS approach could be adopted for use in different countries with varying power-conservation strategies. This is a wrinkle often overlooked during equipment selection, so you will have to make sure that your monitor uses the proper signaling (e.g., DPMS) as well as meets the desired energy-conservation requirements (e.g., Energy Star).

45

The advantage to DPMS signaling is that it can be accomplished with relatively simple programming, such as being integrated into a screen saver. This eliminates the need for specialized video boards, and there are no physical changes to the video port. As an alternative to screen savers, DPMS-compliant video drivers and DPMS-compliant video BIOS are now appearing.

Even Windows 95 screen saver properties have been designed to support DPMS video signaling. However, the video device driver must use either the Advanced Power Management (APM) 1.1 BIOS interface (with support for device 01FFh) or employ the VESA BIOS Extensions for Power Management. You will need to check with the maker of your video board to see if the Windows 95 video drivers support either of these approaches before you can use Windows 95 to manage the monitor's power-saving features.

Power-saving modes are selected based on system inactivity. For example, a system idle for 5 min may set the monitor to its standby mode. After an additional 10 min of inactivity, the monitor will be switched to its suspend mode. After 60 min in suspend mode, the monitor will turn off (where only the monitor's internal microcontroller is running). A keystroke or mouse movement at any point will set the monitor back to its on state (though recovering from a suspend or off state may take 30 to 60 s). In many cases, the actual timing values for each state can be set through the DPMS software.

Nutek

One of the great arguments against DPMS signaling is the need for DPMS-compliant software to properly manipulate the video board's synchronization signals. Sweden has introduced an alternative signaling standard called *Nutek*, which is much simpler and more straightforward than DPMS. Rather than affect horizontal and vertical sync signals, Nutek-compliant monitors simply look for an *absence* of blue video signal. Virtually all modern monitors use three independent analog color signals (red, green, and blue) to generate colors.

When the blue signal is absent, a Nutek-compliant monitor starts an *internal* timing cycle that will gradually step the monitor down to one or more power-saving modes. The moment a screen saver blanks the screen, the Nutek monitor will drop to about 80 percent of its full-power consumption, which is roughly equivalent to the DPMS standby mode. If the blank screen persists for several more minutes, the monitor will drop to its major power-save mode (often down to 10 percent of its full power) where only the CRT heater and microcontroller circuit are active. If the monitor remains in this suspend condition for a prolonged period, the CRT heater will shut down, leaving only the monitor's internal microcontroller active. This off state is typical during idle overnight operation.

Once again, the actual *amount* of power saved at each step will depend on the particular monitor design, so there is no guarantee that a Nutek-compliant monitor is actually Energy Star compliant. That's something you'll have to investigate when shopping for monitors. The timing for each power-down state is usually programmable within the monitor itself.

As you might expect, the Nutek-compliant monitor requires **no** specialized software; any off-the-shelf blank screen saver (or screen image with an absence of blue signal) will work. Whereas a DPMS state is maintained by the DPMS software, the Nutek state is kept track of within the monitor. Since it is the blank screen saver that activated a power down in the first place, any keystroke or mouse movement will cut out the screen saver. When the blue signal resumes, the Nutek-compliant monitor will return to its full-power mode (though there may be a bit of warm-up time for the CRT).

Proprietary techniques

The vast majority of energy-saving monitors in today's marketplace use either DPMS or Nutek signaling, but you may also encounter a variety of proprietary signaling schemes. You are strongly advised to avoid such proprietary monitor approaches simply because technical support, equipment repair, and replacement devices are likely to be difficult or impossible to obtain. Also, there is no promise that proprietary video power-conservation software or drivers will work properly under complex operating systems such as Windows 95.

Configuration and maintenance issues

Like most advances in the PC industry, there are also a series of configuration and maintenance issues that have to be considered. The following points may help you when planning a power-saving video system.

DPMS issues

Display Power Management System has proven to be a popular signaling approach in the PC industry, but it relies on specialized driver software that will manipulate the horizontal and vertical signals of a video board. As a consequence, video drivers and video BIOS become key elements of video power conservation.

□ ***Software compatibility and timing:*** Just as with any other device driver, poorly written DPMS-compliant video drivers can interfere with other system software resulting in system crashes or hang-ups. Check with the video adapter's manufacturer for driver updates and patches. If you rely on DPMS-enabling applications such as After Dark screen savers, disabling or removing the software will disable your power conservation. Time-delay settings for each power state are typically entered into the DPMS screen saver configuration or video driver "properties."

□ ***Swapping video boards and drivers:*** Today, it is virtually impossible to install a new video board without installing a new video driver as well. If another video board does not have a DPMS-compatible driver, you will lose the feature. When shopping for a replacement board, look for the proper drivers, too. If you switch to a generic video driver (e.g., starting Windows 95 in its safe mode), you will also likely lose DPMS capability.

□ ***Swapping monitors:*** Using a non-DPMS monitor in place of a DPMS monitor should not damage anything, and there should be no visual image distortion because the video signals are blanked. However, you will receive little (if any) power savings with a non-DPMS monitor.

Nutek issues

Nutek also relies on software for proper operation, but the demands are not nearly as stringent.

□ ***Screen savers:*** The Nutek approach works best with a blanking screen saver. It does not matter which screen saver you choose as long as there is no blue video signal. However, a regular screen display that contains no blue (a very rare occurrence) will cause a Nutek monitor to power down also. Removing the screen blanker will disable your power conservation. The delay before your screen saver takes control determines when power conservation will begin.

□ ***Monitor timing and programming:*** The time-delay settings for each Nutek power state are programmed into the monitor itself. You will need to refer to the monitor's documentation for default timing and programming options.

The ultimate power conservation

In spite of the advances in power conservation, the age-old question still remains: Should I turn it off or leave it on? Proponents claim that modern systems and peripherals can be turned off safely, but opponents argue that the repeated heating and cooling of system components leads to premature system failure. *Ultimately, there is no logical reason why you should not turn off your PC at the end of a working day.* To understand the answer, you should understand where this debate came from.

When electronic components operate, they dissipate power as heat. The heat causes expansion, which puts stress on the interconnections within an IC, CPU, and other components. Repeatedly turning a system on and off causes repeated heating and cooling of the parts, which eventually fatigue the internal connections. As a result, the part or its interconnections fail. It was argued that by leaving the system on, the parts would reach a stable operating temperature and repeated "thermal fatigue" would be eliminated. This was largely true for older PCs and peripherals which ran hot with high component counts.

Today, low-power operation and efficient IC assembly techniques have dramatically reduced the heat generated in ICs and other components. Just about all of the heat you feel rising from a monitor is from the CRT cathodes, the last remaining throwback to vacuum tubes. While we would not recommend turning a monitor or PC on and off continuously during the day, there is little reason to let the system run all night or all weekend long. Power-conservation modes are great during the working day, but don't be afraid to turn the system off each night. That's the best power conservation there is.

Monitor assembly

Now that you have learned the essential elements of monochrome and color monitors, this part of the chapter shows you how those working elements are implemented in a practical monitor assembly. Figure 2-12 illustrates a typical color computer monitor as you may see it during an actual repair. Keep in mind that internal assemblies will vary a bit between monitors, but Fig. 2-12 provides a good overall example. The first thing you may notice when looking inside a monitor is the layout and position of the printed circuit boards. As a general rule, monitors use three PC board assemblies: one board for the power supply, one board for the video and CRT drive circuits, and one for the raster circuits. However, many monitor assemblies place the power supply circuit directly on the main PC board along with the raster circuits.

The *power supply PC board* is typically a hand-sized assembly that converts ac into several dc voltage levels that will be needed

Exploded view

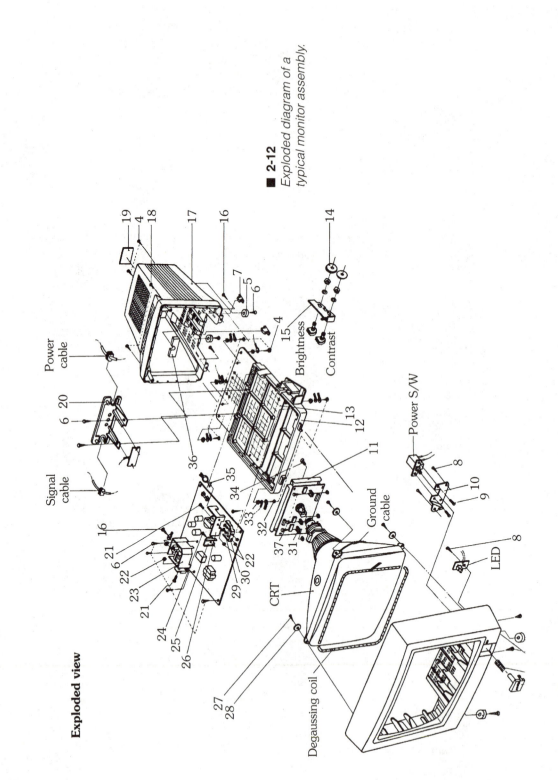

Power cable

Signal cable

Brightness

Contrast

Power S/W

Ground cable

LED

CRT

Degaussing coil

19
4
18
17
16
14
7
5
6
4
15
6 20
6
13
12
11
16
21
6
22
23
21
24
25
26
34 35
33
32
37
31
27
28
36
29 30 22
8
10
9
8

■ **2-12**
Exploded diagram of a typical monitor assembly.

by other monitor circuits. The ac itself may be filtered and fused by a separate small assembly. If there is no stand-alone power supply board in your particular monitor, the supply is probably designed into the main monitor board. As you learned earlier in this chapter, the only voltage that is not produced in the power supply is the high-voltage source. The board is mounted vertically to a metal frame. The metal frame not only provides a rigid mounting platform, but it serves as a chassis common and helps to contain radio-frequency (RF) signals generated by the monitor.

The *CRT drive PCB* attaches directly to the CRT pins through a circular connector. Control (brightness), screen, and focus grid voltages are applied to the CRT through this board. The CRT drive board also usually contains the monochrome or RGB video amplifiers and drivers. Since more video drive circuitry is needed for a color monitor than a monochrome monitor, the CRT drive PC board for a color monitor is usually much larger than that of a monochrome monitor. Once the monitor is unplugged and discharged, make sure that this board is attached evenly and securely to the CRT.

The main *monitor PCB* (often called the *raster board*) contains the vertical raster, horizontal raster, and high-voltage circuits that drive the CRT and direct the electron beam(s) around the screen. Depending on the design of your particular monitor, the main monitor board may contain part or all of the power supply circuit as well. Just about all monitors mount the main PC board to the metal frame horizontally below the CRT neck. This assembly can be difficult to remove because it is obstructed by the CRT neck and yoke as well as the interconnecting wiring that connects to the power supply, front-panel controls, and flyback transformer.

There are some other assemblies in Fig. 2-12 that you should be familiar with. Note the thick wire surrounding the CRT screen. This is known as a *degaussing coil* and is required on all color monitors. Since magnetism influences the path of the electron beams, and all three electron beams must be aligned precisely to achieve proper color, any stray magnetic influence will upset the display's color purity. It is not uncommon for small areas of the metal shadow mask to become magnetized. When this occurs, the color in those magnetized areas will be distorted. The degaussing coil plugs into the power supply circuit. When the monitor is first turned on, a temporary ac voltage in the coil creates a strong alternating magnetic field that acts to clear any magnetized regions in the shadow mask. After a moment, ac is cut off in the coil, and the CRT will operate normally. Chapter 5 provides color purity test and alignment procedures.

The *horizontal output transistor* is shown mounted on a separate heat sink. This is not always the case—the power transistor may be mounted to a heat sink on the main monitor PC board. A horizontal output transistor is a key component that not only drives the horizontal deflection yoke but the flyback transformer as well. You may also notice the *high-voltage anode* which originates at the flyback transformer and is little more than a metal prong covered by a large red plastic insulator. This anode must be inserted fully in the CRT. **Never** *touch the anode wire or connector (even when the monitor is turned off)*. This conductor is carrying 15 kV or higher, so a *very* dangerous shock hazard exists. The monitor must be *unplugged* and the CRT must be safely discharged before working with the high-voltage anode. You will learn how to discharge the high-voltage anode later in this chapter.

Finally, you should take note of any metal shrouds or coverings that are included in a monitor. Metal shielding serves two very important purposes. First, the oscillators and amplifiers in a computer monitor produce RF signals that have the potential to interfere with radio and TV reception. The presence of metal shields or screens helps to attenuate any such interference, so always make it a point to replace shields securely before testing or operating the monitor. Second, large CRTs (larger than 17 in.) use very high voltages (25 kV or more) at the anode. With such high potentials, X radiation becomes a serious concern. Cathode-ray tubes with lower anode voltages can usually contain X rays with lead in the CRT glass. Metal shields are added to the larger CRTs in order to stop X rays from escaping the monitor enclosure. When X-ray shielding is removed, it is vital that it be replaced before the monitor is tested and returned to service. X-ray shields will usually be clearly marked when you remove the monitor's rear cover.

Warnings, cautions, and human factors

Please read this section carefully! One of the most important aspects of all computer service is *safety* for you and those working and living around you. Computer monitors present a special set of safety hazards that most other PC equipment does not, so it is important that you read and understand the following safety information. *If you are uncomfortable working around such hazards or* *do not understand the implications of these warnings, **do not** attempt to repair the monitor. Refer the service to someone else.*

AC voltages

The ac voltage available from any wall receptacle can be very dangerous if you touch a live wire or connection. No matter how harmless your monitor may appear, always remember that potential shock hazards do exist. Once the monitor is disassembled, there can be several locations (generally around the power supply and ac filter) where live ac voltage is exposed and easily accessible. Domestic electronic equipment operates from 120 V ac at 60 Hz. Some European countries use 240 V ac at 50 Hz. When this kind of voltage potential establishes a path through your body, it causes a flow of current that may be large enough to stop your heart. Since it only takes about 100 mA to trigger a cardiac arrest, and a typical monitor fuse is rated for 1 or 2 A, fuses and circuit breakers will **not** protect you.

It is your skin's resistance that limits the flow of current through the body. Ohm's law states that, for any voltage, current flow increases as resistance drops (and vice versa). Dry skin exhibits a high resistance of several hundred thousand ohms, whereas moist, cut, or wet skin resistance can drop to only several hundred ohms. This means that even comparatively low voltages can produce a shock if your skin resistance is low enough. Some examples will help to demonstrate this action.

Suppose your hands come across a live 120 V ac circuit. If your skin is dry (say, 120 kΩ), you would experience an electrical shock of 1mA (120 V ac/120,000 Ω). The result would be a harmless, probably brief, tingling sensation. But after a hard day's work, perspiration can decrease your skin's resistance (perhaps 12 kΩ). This would allow a far more substantial shock of 10 mA (120 V ac/12,000 Ω). At that level, the shock can paralyze you and make it difficult or impossible to let go of the "live" conductors. A burn (perhaps serious) could result at the points of contact, but it probably would not be fatal. Now consider what happens if your hands or clothing are wet. Effective skin resistance can drop very low (e.g., 1.2 kΩ). At 120 V, the resulting shock of 100 mA (120 V ac/1,200 Ω) would often be instantly fatal without immediate CPR. *Take the following steps to protect yourself from injury.*

1. Keep the monitor *unplugged* (not just turned off) as much as possible during disassembly and repair. When you *must* perform a service procedure that requires power to be applied,

53

plug in the monitor just long enough to perform your procedure and then unplug it again. This makes the monitor safer for you, as well as your co-worker(s), spouse, children, and pets that may happen along. For added safety, plug in your monitor through an *isolation transformer* such as the one in Fig. 2-13.

■ **2-13** *A B+K Model 1604 isolated ac power supply (B+K Precision)*

2. Whenever you must work on a monitor's power supply, try to wear rubber gloves. These will insulate your hands just like insulation on a wire. You may think that rubber gloves are inconvenient and uncomfortable, but they are far better than the inconvenience and discomfort of an electric shock. Make it a point to wear a long-sleeved shirt with sleeves rolled down to insulate your forearms.

3. If rubber gloves are absolutely out of the question for one reason or another, remove **all** metal jewelry and work with one hand behind your back. The metals in your jewelry are excellent conductors. Should your ring or watchband hook onto a "live" ac line, it can conduct current directly to your skin. By keeping one hand behind your back, you cannot grasp both ends of a live ac line to complete a strong current path through your heart.

4. Work dry! **Do not** work with wet hands or clothing. **Do not** work in wet or damp environments. Make sure that any available fire extinguishing equipment is suitable for electrical fires.

5. Treat electricity with tremendous respect. Whenever electronic circuitry is exposed (especially power supply

circuitry), a shock hazard **does** exist. Remember that it is the flow of current through your body, not the voltage potential, that can injure you. Insulate yourself as much as possible from any exposed wiring.

High voltages

The high anode voltages available in a monitor also present a serious shock hazard. Most monitors can produce voltages easily exceeding 15,000 V. Fortunately, high-voltage power supplies are not designed to source significant current, but serious burns can be delivered with ease. Not only is there a great risk of injury, but normal test probes (e.g., multimeter test leads) only provide insulation to about 600 V. Testing high voltages with standard test leads could electrocute you right through the lead's insulation! *Be sure to use specially designed high-voltage probes when measuring CRT anode voltages.*

Even after the monitor is turned off and unplugged, the CRT bell (its large flat face) can retain a significant charge, so you must discharge the CRT anode before working on the monitor. You will also need to discharge the monitor after each time it is powered up and tested on your workbench. Make sure the monitor is turned off. Use a screwdriver as shown in Fig. 2-14. Attach a large alligator clip from the screwdriver to the monitor's metal frame. Slide the

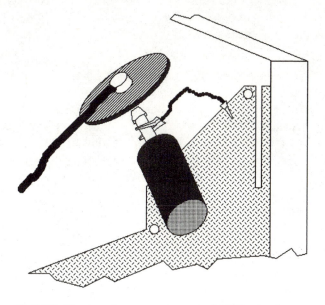

■ **2-14** *Discharging an unpowered CRT prior to servicing.*

screwdriver blade *gently* under the anode insulator and contact the metal anode clip. This effectively shorts any potential on the CRT to ground. You will hear mild crackling sounds as the CRT is discharged. *Be very careful to avoid twisting the screwdriver inside the CRT—it is still a fragile glass assembly.* Short the CRT for several seconds to ensure a full discharge.

X radiation

When an electron moving at high speed strikes a phosphor particle, a number of X-ray particles are liberated. The CRT face is basically a big X-ray emitter. Since the early days of television, when X rays were discovered from TV picture tubes, ordinary lead was added to the CRT glass. Monochrome monitor CRTs also use leaded glass (usually a barium-lead combination) to contain X-ray particles and are typically effective up to anode voltages of 18 kV. Color CRTs are more likely to use a barium-strontium combination (instead of lead), which can provide protection up to 30 kV or so. Monitor glass is quite effective at containing X rays as long as the high-voltage level at the anode is not too high. If high voltage does become too high, X-ray generation increases and ultimately browns the glass (more specifically the lead in the glass) and escapes the CRT. Since long-term exposure to X radiation has been linked to cancer and other health problems, it is important for you to check and correct excessive high-voltage conditions in computer monitors. This helps to ensure your safety as well as the long-term safety of the monitor user(s).

There are actually three parts to a CRT: the *panel* (or front face), the *funnel* (or bell), and the *neck*. Each part of the CRT is manufactured separately and sealed together during the final manufacturing process. This means that all three parts of the CRT are typically three different glass formulas. You should always remember that while *all* CRT glass is resistant to X rays, protection is typically concentrated in the panel (front) glass. This raises another safety question when working *behind* the CRT during procedures such as alignment. Ultimately, it should be perfectly safe to work behind a CRT as long as its high-voltage level is not excessive. When high voltage is excessive, X rays will typically begin to penetrate the neck and funnel before penetrating the panel. When you consider the large number of computer users who work all day facing each other (looking at the back end of each other's monitors), you can understand how important it is for you to ensure safe monitor operation. The following points will help you deal with X-ray dangers.

1. X rays are *always* generated in CRT operation, but protected glass used in CRTs will contain X rays as long as the high voltage level at the anode is within its specified limits. If high voltage becomes excessive, X rays may penetrate the glass. One of your first routine checks when a monitor is opened for service should be to check the high-voltage level with a *calibrated* high-voltage probe. If high voltage exceeds the monitor's specifications, adjust the high voltage or troubleshoot the high-voltage circuit. Under **no** circumstances should the monitor be returned to service with excessive high voltage.

2. If you do discover that high voltage is excessive, do not operate the monitor any longer than *absolutely necessary* while troubleshooting the cause of the excessive high voltage.

3. Larger CRTs using high voltages in excess of 25 kV (where the chances of excessive high voltage are much greater) often employ a set of supplemental radiation shields bolted around the CRT funnel and neck. **Do not** test or operate the monitor with these supplemental shields removed, especially if you are aware that high voltage is excessive.

4. If the CRT must be replaced during the course of your repair, replace the CRT with only an *exact* replacement part. A different part may not offer enough lead in the glass. There may also be variations in the phosphors quality and characteristics that will affect the ultimate quality of the display. Similarly, if you replace components in the high-voltage circuit, be sure to use only exact replacement parts. This ensures that the high-voltage circuit will behave as expected.

Carrying the CRT

Ultimately, the CRT is still a glass tube. It can crack and shatter like any other large glass vessel. When replacing the CRT, there are some common-sense things to keep in mind. First, wear gloves and goggles when carrying a CRT (especially a large CRT). **Never** under any circumstances should you lift, suspend, or carry the CRT by its thin neck. Instead, hold the CRT with its face to your chest and grasp it around the width of its funnel. When setting the CRT down, **do not** rest it on the neck. Rest the CRT face down on a thick, soft cushion such as plush towels or thick foam padding. A soft resting place prevents the CRT screen from being scratched. Also make sure that the resting place is away from where people and pets walk and is very stable (e.g., a table top). Leaving the CRT on a chair or the floor can result in accidental injury (and ir-

57

reparable damage to the CRT). Finally, remember that CRTs are typically heavy items (especially large CRTs). When lifting CRTs, be sure to lift "from the knees" to reduce the stress on your back. It's tough to fix a monitor while you're in traction.

CRT troubleshooting

Now that you have some insights into the CRT and CRT-based monitor basics, this part of the chapter covers some common monitor troubleshooting procedures which are often related to CRT problems. Keep in mind that there may be more complex electronic causes for these problems, and these are covered in detail later in the book.

When considering a CRT replacement, you should remember that the CRT is typically the most expensive part of the monitor. For larger monitors, the CRT becomes an even larger percentage of the monitor's overall cost. In many cases, the cost for a replacement CRT approaches the original cost of the entire monitor. As a consequence, you should carefully evaluate the *economics* of replacing the CRT versus buying a new monitor outright.

Symptoms and solutions

Symptom 1 *The screen images are dim or dark.* Before getting into this type of symptom, you should be sure that the brightness and contrast controls have not been turned down. Generally speaking, the image should have acceptable brightness with the brightness and contrast controls in their detent (50 percent) positions. If the brightness and contrast are significantly above detent and the image still appears dark, there may be a more serious problem with the monitor. Also make it a point to check for other factors such as excessive dust on the CRT or the presence of a glare shield that may artificially reduce the image's brightness.

Once you have checked these basics, you should eliminate any external factors driving the CRT. First, measure your anode voltage at the CRT and your 135-V output (usually referred to as the B+ output) from the power supply. A low high-voltage level will result in low brightness. Simply, if the B+ supply checks good, but anode voltage is low, you have a problem in the monitor's high-voltage circuit (refer to Chapters 7 and 8 for more detailed circuit information). If the B+ supply is low (causing the low anode voltage), you will need to troubleshoot your power supply (refer to Chapter 6).

Next, check the 6.5-V output from the power supply. This is typically used to run the CRT's filament(s). Those glowing cathodes are the source of the electrons that strike the phosphors. If the filament voltage is low, there may be fewer electrons being produced at the CRT cathode, and the resulting image is dimmer. If the filament voltage is low, repair the power supply.

Also check the midrange voltages generated by the power supply (i.e., 12 and 20 V). These voltages are used by the monitor's raster and video circuits. If either of these voltages are low, the video amplifiers may not be able to develop their full signal power resulting in a dim image. Repair or replace the defective power supply.

Ultimately, if your voltages measure correctly, your CRT may have one or more internal faults. Like most vacuum tubes, CRTs typically degrade and fail gradually over the course of years. Corrosion and shorting in its internal grids and guns can easily account for diminished output. If you have a CRT checker-rejuvenator, test the CRT and correct internal problems if possible. Otherwise, your only alternative may be to replace the CRT.

Symptom 2 *The image contains dark blacks and overdriven whites.* As with the first symptom, you may be faced with a defective power supply, particularly the 12 or 20 V outputs. If these outputs are weak, the video amplifier circuits may become unstable resulting in a nonlinear video output. If the power supply is running properly, you may have weak color guns in the CRT that are causing a nonlinear output (bad gamma). If you have a CRT tester-rejuvenator, you can check the CRT for bad guns (and perhaps zap some new life into them). If the guns read weak, but you cannot rejuvenate them, you should consider replacing the CRT.

Symptom 3 *The monitor displays poor colors or a nonlinear gray scale.* The image may contain an excessive amount of red, green, or blue. This may be due to a weakness in one or more video amplifiers. With a pure white image being displayed, use an oscilloscope and check the signal output at each video amplifier. If you find an unusually low output, you may be able to boost the output with an adjustment on the video amplifier board. Otherwise, you will have to troubleshoot or replace the video amplifier board. If all three outputs are equal, your fault is likely in the CRT itself.

Cathode-ray tube phosphors age normally regardless of use and will age even faster when used in normal operation. Eventually, the phosphors will "burn out" resulting in a poor, dull representation of colors. This is similar to, but not as dramatic as, phosphor "burn in" where latent images remain permanently marked on the screen.

Symptom 4 *The raster appears unusually bright and may appear colored.* In most cases, there is a short inside the CRT, probably between two color guns. It is also possible that the CRT's control grid may be opened (allowing excessive screen brightness). If you have a CRT tester-rejuvenator, you can check the CRT for internal faults. If a fault is detected but cannot be fixed, the CRT will have to be replaced. If the CRT checks properly, you should suspect a fault or short circuit in the video amplifier board. Unfortunately, short circuits are very difficult to spot, so your best course is usually to replace the video amplifier board outright.

Knowing the tools

3

TROUBLESHOOTING TAKES SOME AMOUNT OF TEST EQUIP-ment. Test equipment allows you to measure important circuit parameters such as voltage, current, resistance, capacitance, and semiconductor junction conditions. Additional test equipment can let you follow logic conditions and view complex analog waveforms at critical points in the circuit. Of course, there is also specialized equipment designed exclusively for testing computer monitors. This chapter introduces you to the purposes and applications of a variety of test equipment.

Before we get into this chapter, you should understand what test instruments do. Instruments are *sensors* that let you peek into the operations of a circuit. The more sensors you have, the better those sensors are, and the more skill or experience you have with them, the better your determinations of a problem will be. You will need instruments if you plan to troubleshoot a monitor to the component level because you simply cannot discern the condition of individual components any other way. Ideally, a well-equipped test bench in the hands of an experienced technician can track a problem to a specific component and replace the relatively inexpensive component in a short period of time. Experienced technicians and long-time electronics enthusiasts will probably have some of this equipment already on hand.

However, many readers do not have test equipment. This book does not suggest that you run out and invest thousands of dollars to purchase test equipment to fix a monitor that is only worth several hundred dollars. That makes no economic sense. You can troubleshoot *symptomatically* by replacing subassemblies based on the symptoms that the monitor is showing. Symptomatic troubleshooting is employed by many major repair houses where labor costs and repair volumes are too high to spend time tracking down bad components—just swap the board where the problem most likely is located. The subassembly certainly costs more than individual components, but the time and equipment needed to do the repair are a lot less. This book is written to address both component-level and symptomatic troubleshooting wherever possible. If you have test equipment available, or have considered purchasing some, this chapter will help you find what equipment you need and how it can help you.

Tools and materials

If you don't have a well-stocked toolbox, now is a good time to consider the tools and materials you need. Before you begin a repair, gather a set of small hand tools and some inexpensive materials. *Never* underestimate the value of having the proper tools; they can often make or break your repair efforts.

Hand tools

Hand tools are basically used to disassemble and reassemble your monitor's housings and enclosures. It is not necessary to stock top-quality tools, but your tools should be of the proper size and shape to do the job. Since most monitors are relatively spacious devices, you can usually do well with small to medium-sized hand tools.

Screwdrivers should be the first items on your list. Most monitor assemblies are held together with small or medium-sized Phillips-type screws. Once you are able to remove the outer housings, you will probably find that most other internal parts are also held in place with Phillips screws. Consider obtaining one or two medium Phillips screwdrivers as well as one small version. You will almost never need a large screwdriver. Each screwdriver should be about 10.16 cm (4 in.) to 15.24 cm (6 in.) long with a wide handle for a good grip. Round out your selection of screwdrivers by adding one small and one medium regular (flat blade) screwdriver. You won't use them as often as Phillips screwdrivers, but regular screwdrivers **will** come in handy.

There are three specialized types of screw heads that you should be aware of. *Allen* screws use a hex (six-sided) hole instead of the regular or Phillips-type slots. *Torx* and *spline* screws use specially shaped holes that only accept the corresponding size and shape of driver. It is a good idea to keep a set of small hex keys on hand, but you will rarely find specialized screw heads in today's monitors. Torx and spline screws are only rarely encountered.

Wrenches are used to hold hex-shaped bolt heads or nuts. There are not many instances when you need to remove nuts and bolts, but an inexpensive set of small electronics-grade open-ended wrenches is recommended. If you prefer, a very small adjustable wrench can be used instead. Wrenches are handiest when removing power transistors from their heat sinks or mounting assemblies.

Needlenose pliers are valued additions to your toolbox. Not as bulky and awkward as ordinary mechanic's pliers, needlenose pliers can be used to grip or bend both mechanical and electronic parts. Needlenose pliers can also serve as heat sinks during desoldering or soldering operations. Obtain a short nose and long nose set of these pliers. Short nose pliers make great heat sinks and can grasp parts securely. Long nose pliers are excellent for picking up and grasping parts lost in a monitor assembly. All sets of needlenose pliers should be small, good-quality electronics-grade tools with well-insulated padding on the handles.

Diagonal cutters are also an important part of your tool collection. Cutters are used to cut wire and component leads when working with a monitor's electronics. You really only need one good set of cutters, but the cutters should be small, good-quality electronics-grade tools with well-insulated padding on the handles. Cutters should also have a low profile and a small cutting head to fit in tight spaces. *Never* use cutters to cut plastic, metal, or PC board material.

Add a pair of tweezers to your tool kit. The tweezers should be small, long, and made from antistatic plastic material. Metal tweezers should be avoided wherever possible to prevent accidental short circuits (as well as a shock hazard) if they come into contact with operating circuitry. Metal tweezers can also conduct potentially damaging static charges into sensitive integrated circuits (ICs).

Soldering tools

You need at least one good-quality, general purpose soldering iron to repair your monitor's circuitry. A low-wattage (12 to 20 W) iron with a fine tip is usually best (Fig. 3-1). You can obtain a decent soldering iron from any local electronics store. Most soldering irons are powered directly from ac, and these are just fine for general touch-ups and heavier work. However, you should consider a dc-powered or gas-fueled iron for desoldering delicate, static-sensitive ICs. No matter what iron you buy, try to ensure that it is recommended as "static-safe."

The iron ABSOLUTELY must have its own metal stand! *Never*, under *any* circumstances, allow a soldering iron to rest on a counter or table top unattended. The potential for nasty burns or

63

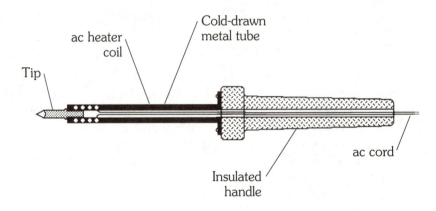

3-1 *Cross-section of a basic soldering pencil.*

fire is simply too great. Keep a wet sponge handy to periodically wipe the iron's tip. Invest in a roll of good-quality electronics-grade rosin-core solder.

Desoldering tools are necessary to remove faulty components and wires. Once the solder joint is heated with the soldering iron, a desoldering tool can remove the molten solder to free the joint. A solder vacuum uses a small spring-loaded plunger mounted in a narrow cylinder. When triggered, the plunger recoils and generates a vacuum which draws up any molten solder in the vicinity. Solder wick is little more than a fine copper braid. By heating the braid against a solder joint, molten solder wicks up into the braid through capillary action. Such conventional desoldering tools are most effective on through-hole components.

Surface-mounted components can also be desoldered with conventional desoldering tools, but there are more efficient techniques for surface-mount parts. Specially shaped desoldering tips can ease surface-mount desoldering by heating all of the component's leads simultaneously (Fig. 3-2). Powered vacuum pumps can also be used to remove molten solder much more thoroughly than spring-loaded versions.

Multimeters

Multimeters are by far the handiest and most versatile pieces of test equipment that you will ever use (Fig. 3-3). If your toolbox does not contain a good-quality multimeter already, now would be a good time to consider purchasing one. Even the most basic digital multimeters are capable of measuring resistance, ac and dc voltage, and ac and dc current. For under $150, you can buy a dig-

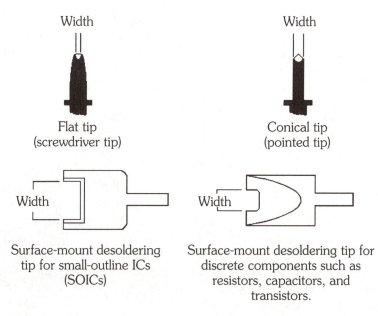

Width   Width

Flat tip
(screwdriver tip)

Conical tip
(pointed tip)

Width   Width

Surface-mount desoldering
tip for small-outline ICs
(SOICs)

Surface-mount desoldering tip for
discrete components such as
resistors, capacitors, and
transistors.

■ **3-2** *A selection of basic surface-mount soldering/desoldering tips.*

■ **3-3**
*A B+K Model 390 Test Bench
multimeter. (B+K Precision)*

ital multimeter that includes handy features like a capacitance checker, a frequency meter, an extended current measuring range, a continuity buzzer, and even a diode and transistor checker. These are features that will aid you not only in monitor repairs, but in many other types of electronic repairs as well. Digital multimeters are easier to read, more tolerant of operator error, and more precise than their analog predecessors.

Meter setup

For most multimeters, there are only three considerations during setup and use. First, turn the meter on. Unlike analog multimeters, digital multimeters require power to operate liquid crystal or light-emitting diode displays. Make sure that you turn meter power off again when you are finished with your testing. Power awareness will help you conserve the meter's battery life. Second, your meter must be set to its desired function or mode. The *function* may be frequency, voltage, capacitance, resistance, and so on, depending on the particular physical parameter that you wish to measure. Finally, you must select the meter's *range* for its selected function. Ideally, you should choose the range that is nearest to (but above) the level you expect to measure. For example, suppose you are measuring a 9-V dc transistor battery. You would set your meter to the dc voltage function and then set your range as close to (but greater than) 9 V dc as possible. If your voltage ranges are 0.2 V dc, 2 V dc, 20 V dc, and 200 V dc, the 20-V dc range is your best choice.

If you are unsure about just which range to use, start by choosing the *highest* possible range. Once you actually take some measurements and get a better idea of the actual reading, you can then adjust the meter's range "on the fly" to achieve a more precise reading. If your reading exceeds the meter's current range, an *overrange* warning will be displayed until you increase the meter's range above the measured value. Some digital multimeters are capable of automatically selecting the appropriate range setting (called *autoranging*) once a signal is applied.

Checking test leads

It is usually a good idea to check the integrity of your test leads from time to time. Since test leads undergo a serious amount of tugging and general abuse, you should be able to confirm that the probes are working as expected. There are few experiences more frustrating than to replace parts that your meter suggested were faulty only to discover that the meter leads had an internal fault.

66

To check your probes, set your meter to the *resistance* function and then select the *lowest* scale (i.e., 0.1 Ω). You will see an *over-range* condition, which is expected when setting up for resistance measurements. Check to be sure that both test probes are inserted into the meter properly; then touch the probe tips together. The resistance reading should drop to about 0 Ω to indicate that your meter probes are intact. If you do not see roughly 0 Ω, check your probes carefully. After you have proven out your test probes, return the multimeter to its original function and range so that you may continue testing.

You may see the terms static and dynamic related to multimeter testing. *Static* tests are usually made on components (either in or out of a circuit) with power removed. Resistance, capacitance, and semiconductor junction tests are all static tests. *Dynamic* tests typically examine circuit conditions, so power must be applied to the circuit, and all components must be in place. Voltage, current, and frequency are the most common dynamic tests. Use extreme caution when making dynamic tests. Alternating current and high voltage are present in computer monitors, and both represent serious safety hazards. If you have not read the section entitled *Warnings, Cautions, and Human Factors* in Chapter 2, please do so now. **Do *not*** *begin testing monitors until you understand the dangers involved.*

Measuring voltage

Every signal in your monitor has a certain amount of voltage associated with it. By measuring signal voltages with a multimeter (or other test instrument), you can usually make a determination as to whether or not the signal is correct. Supply voltages that provide power to your circuits can also be measured to ensure that components are receiving enough energy to operate. Voltage tests are the most fundamental (and the most important) dynamic tests in electronic troubleshooting. **Do not** use your multimeter to test voltages over 600 V (or the input limit of the multimeter, whichever is lower).

Multimeters can measure both dc voltages (marked DCV or Vdc) and ac voltages (marked ACV or Vac) directly. Remember that *all* voltage measurements are taken *in parallel* with the desired circuit or component. *Never* interrupt a circuit and attempt to measure voltage in series with other components. Any such reading would be meaningless, and your circuit will probably not even function.

Follow your setup guidelines and configure your meter to measure ac or dc voltage as required and then select the proper range for the voltages you will be measuring. If you are unsure just what range to use, always start with the *largest* possible range. An autoranging multimeter will set its own range once a signal is applied. Place your test leads across (*in parallel* with) the circuit or part under test (PUT) as shown in Fig. 3-4; then read voltage directly from the meter's digital display. Direct current voltage readings are polarity sensitive, so if you read +5 V dc and reverse the test leads, you will see a reading of −5 V dc. Alternating current voltage readings are not polarity sensitive.

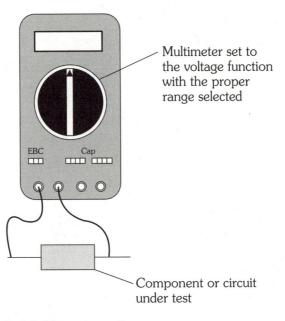

Multimeter set to the voltage function with the proper range selected

Component or circuit under test

■ **3-4** *Measuring voltage.*

Measuring current

Most general purpose multimeters allow you to measure alternating current (marked ACA or Iac) and direct current (marked DCA or Idc) in an operating circuit, although there are typically fewer ranges to choose from. As with voltage measurements, current is a dynamic test, so the circuit or component being tested must be under power. However, current *must* be measured *in series* with a circuit or component.

Unfortunately, inserting a meter in series is not always a simple task. In many cases, you must interrupt a circuit at the point you wish to measure and then connect your test leads across the break. While it may be quite easy to interrupt a circuit, remember that you must also put the circuit back together, so use care when choosing a point to break. *Never* attempt to measure current in parallel across a component or circuit. Current meters, by their very nature, exhibit a very low resistance across their test leads (often below 0.1 Ω). Placing a current meter in parallel can cause a short circuit across a component that can damage the component, the circuit under test, or the meter itself.

Set your multimeter to the desired function (DCA or ACA) and select the appropriate range. If you are unsure about the proper range, set the meter to its *largest* possible range. It is usually necessary to plug your positive test lead into a "current input" jack on the multimeter. Unless your multimeter is protected by an internal fuse (most meters *are* protected, but review your particular meter to be sure), its internal current measurement circuits can be damaged by excessive current. Make sure that your meter can handle the maximum amount of current you are expecting.

Turn off all power to a circuit before inserting a current meter. Deactivation prevents any unpredictable or undesirable circuit operation when you actually interrupt the circuit. If you wish to measure power supply current feeding a circuit such as in Fig. 3-5, break the power supply line at any convenient point, insert the meter carefully, and reapply power. Read current directly from the meter's display. This procedure can also be used for taking current measurements within a circuit. Remember to deactivate the circuit, remove the meter, and reconnect the interrupted line before continuing.

Measuring resistance

Resistance (ohms) is the most common static measurement that your multimeter is capable of. This is a handy function not only for checking resistors themselves but for checking other resistive elements like wires, solenoids, motors, connectors, and some basic semiconductor components. Resistance is a static test, so all power to the component or circuit must be removed. It is usually necessary to remove at least one component lead from the circuit to prevent interconnections with other components from causing false readings.

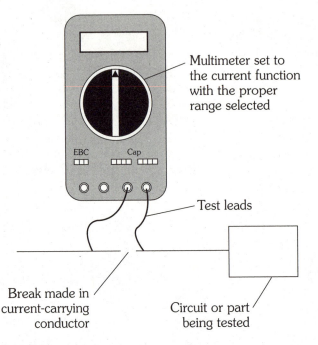

Multimeter set to
the current function
with the proper
range selected

EBC   Cap

Test leads

Break made in
current-carrying
conductor

Circuit or part
being tested

■ **3-5** *Measuring current.*

70

Ordinary resistors, coils, and wires can be checked simply by switching to a resistance function (often marked ohms or with the Greek symbol omega, Ω) and selecting the appropriate range. Autoranging multimeters will select the proper range after the meter's test leads are connected. Many multimeters can reliably measure resistance up to about 20 MΩ. Place your test leads *in parallel* across the component as shown in Fig. 3-6 and read resistance directly from the meter's display. If resistance exceeds the selected range, the display will indicate an overrange (or infinite resistance) connection.

Continuity checks are made to ensure a reliable low-resistance connection between two points. For example, you could check the continuity of a cable between two connectors to ensure that both ends are connected properly. Set your multimeter to a low-resistance scale; then place your test leads across both points to measure. Ideally, good continuity should be about 0 Ω. Continuity tests can also be taken to show that a short circuit has *not* occurred between two points

Measuring capacitors

The two methods of checking a capacitor using your multimeter are by exact measurement and by a quality check. The *exact measure-*

Knowing the tools

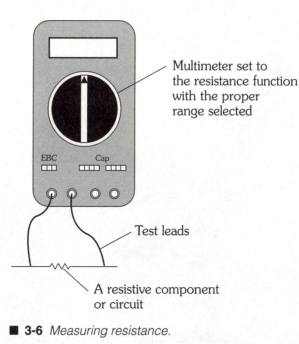

Multimeter set to the resistance function with the proper range selected

EBC

Cap

Test leads

A resistive component or circuit

■ **3-6** *Measuring resistance.*

ment test determines the actual value of a capacitor. If the reading is close enough to the value marked on the capacitor, you know that the device is good. If not, you know the device is faulty and should be replaced. Exact measurement requires your multimeter to be equipped with a built-in capacitance checker. If your meter does not have a built-in capacitance checker, you can measure a capacitor directly on any other type of specialized component checker such as the one shown in Fig. 3-7. You could also use your multimeter to perform a simple *quality check* of a suspect capacitor.

Capacitor checkers, whether built into your multimeter or part of a stand-alone component checker, are extremely simple to use. Turn off all circuit power. Set the function to measure capacitors, select the range of capacitance to be measured, and place your test probes *in parallel* across the capacitor to be measured. You should remove at least one of the capacitor's leads from the circuit being tested to prevent the interconnections of other components from adversely affecting the capacitance reading. In some cases, it may be easier to remove the suspect part entirely before measuring it. Some meters provide test slots that let you insert the component directly into the meter's face. Once in place, you can read the capacitor's value directly from the meter display.

If your multimeter is not equipped with an internal capacitor checker, you could still use the resistance ranges of your ohm-

■ **3-7**
A B+K Model 815 Parts Tester. (B+K Precision)

meter to approximate a capacitor's quality. This type of check provides a "quick and dirty" judgment of whether the capacitor is good or bad. The principle behind this type of check is simple; all ohmmeter ranges use an internal battery to supply current to the component under test. When that current is applied to a working capacitor as shown in Fig. 3-8, it will cause the capacitor to charge. Charge accumulates as the ohmmeter is left connected. When first connected, the uncharged capacitor draws a healthy

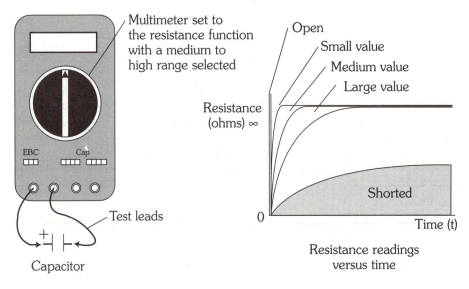

Multimeter set to the resistance function with a medium to high range selected

Open
Small value
Medium value
Large value

Resistance (ohms) ∞

Shorted

0

Time (t)

EBC   Cap

Test leads

Capacitor

Resistance readings versus time

■ **3-8** *Measuring the quality of a capacitor using the resistance function.*

amount of current, and this reads as low resistance. As the capacitor charges, its rate of charge slows down and less and less current is drawn as time goes on, which results in a gradually increasing resistance level. Ideally, a fully charged capacitor stops drawing current, resulting in an overrange or infinite resistance display. When a capacitor behaves in this way, it is probably good.

Understand that you are not actually measuring resistance *or* capacitance here, but only the profile of a capacitor's charging characteristic. If the capacitor is extremely small (in the picofarad range) or is open circuited, it will not accept any substantial charge, so the multimeter will read infinity almost immediately. If a capacitor is partially (or totally) short circuited, it will not hold a charge, so you may read 0 Ω, or resistance may climb to some value below infinity and remain there. In either case, the capacitor is probably defective. If you doubt your readings, check several other capacitors of the same value and compare readings. Be sure to make this test on a moderate- to high-resistance scale. A low-resistance scale may overrange too quickly to achieve a clear reading.

Checking diodes

Many multimeters provide a special "diode" resistance scale used to check the static resistance of common diode junctions. Since working diodes only conduct current in one direction, the diode check lets you determine whether a diode is open or short circuited. Remember that diode checking is a static test, so all power must be removed from the part under test. Before making measurements, be certain that at least one of the diode's leads has been removed from the circuit. Isolating the diode prevents interconnections with other circuit components from causing false readings.

Select the diode option from your multimeter's resistance functions. You generally do not have to bother with a range setting while in the diode mode. Connect your test leads *in parallel* across the diode in the forward-bias direction as shown in Fig. 3-9. A working silicon diode should exhibit a static resistance between about 450 and 700 Ω, which will read directly on the meter's display. Reverse the orientation of your test probes to reverse bias the diode as in Fig. 3-10. Since a working diode will not conduct at all in the reverse direction, you should read infinite resistance.

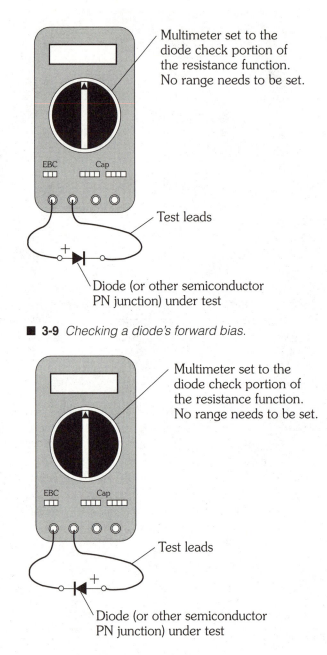

Multimeter set to the
diode check portion of
the resistance function.
No range needs to be set.

Test leads

Diode (or other semiconductor
PN junction) under test

■ **3-9** *Checking a diode's forward bias.*

Multimeter set to the
diode check portion of
the resistance function.
No range needs to be set.

Test leads

Diode (or other semiconductor
PN junction) under test

■ **3-10** *Making a reverse-bias diode check.*

A short-circuited diode will exhibit a very low resistance in the for-
ward- *and* reverse-biased directions. This indicates a shorted
semiconductor junction. An open-circuited diode will exhibit very
high resistance (usually infinity) in both its forward- and reverse-
biased directions. A diode that is opened or shorted must be re-

placed. If you feel unsure of how to interpret your measurements, test several other comparable diodes and compare readings.

Checking transistors

Transistors are slightly more sophisticated semiconductor devices that can be tested using a transistor checking function on your multimeter or component checker. Transistor junctions can also be checked using a multimeter's diode function. The following procedures show you both methods of transistor checking.

Some multimeters feature a built-in transistor checker that measures a transistor's gain (called β, *beta*, or hf_e) directly. By comparing measured gain to the gain value specified in manufacturer's data (or measurements taken from other identical parts), you can easily determine whether the transistor is operating properly. Multimeters with a transistor checker generally offer a test fixture right on the meter's face. The fixture consists of two 3-hole sockets: one socket for NPN devices and another socket for PNP devices. If your meter offers a transistor checker, insert the transistor into the test fixture on the meter's face.

Since all bipolar transistors are three-terminal devices (emitter, base, collector), they must be inserted into the meter in their proper lead orientation before you can achieve a correct reading. Manufacturer's data sheets for a transistor will identify each lead and tell you the approximate gain reading that you should expect to see. Once the transistor is inserted appropriately in its correct socket, you can read gain directly from the meter's display.

Set the meter to its transistor checker function. You should not have to worry about selecting a range when checking transistors. Insert the transistor into its test fixture. An unusually low reading (or zero) suggests a short-circuited transistor, whereas a high (or infinite) reading indicates an open-circuited transistor. In either case, the transistor is probably defective and should be replaced. If you are uncertain, test several other identical transistors and compare your readings.

If your particular multimeter or parts tester only offers a diode checker, you can approximate the transistor's condition by measuring its semiconductor junctions individually. Figure 3-11 illustrates the transistor junction test method. Although structurally different from conventional diodes, the base-emitter and base-collector junctions of bipolar transistors behave just like diodes. As a general rule, you should remove the transistor from its circuit to prevent false readings caused by other interconnected components. Junction testing is also handy for all varieties of surface-mount transistors that will not fit into conventional multimeter test sockets.

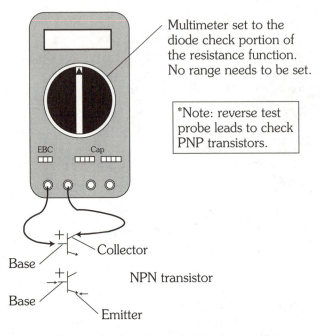

Multimeter set to the diode check portion of the resistance function. No range needs to be set.

*Note: reverse test probe leads to check PNP transistors.

EBC   Cap

Base

+ Collector

Base

+ NPN transistor

Emitter

■ **3-11** *Testing the junctions of a bipolar transistor.*

Set your multimeter to its diode resistance function. If your suspect transistor is NPN type (manufacturer's data or a corresponding schematic symbol will tell you), place your *positive* test lead at the transistor's base and place your *negative* test lead on the transistor's emitter. This test lead arrangement should forward bias the transistor's base-emitter junction and result in a normal amount of diode resistance (450 to 700 Ω). Reverse your test leads across the base-emitter junction. The junction should now be reverse biased showing infinite resistance. Repeat this entire procedure for the base-collector junction.

If your suspect transistor is the PNP type, the placement of your test leads will have to be reversed from the procedure just described. In other words, a junction that is forward biased in an NPN transistor will be reverse biased in a PNP device. To forward bias the base-emitter junction of a PNP transistor, place your positive test lead on the emitter and your negative test lead on the base. The same concept holds true for the base-collector junction.

Once both junctions are checked, measure the diode resistance from collector to emitter. You should read infinite resistance in *both* test lead orientations. Although there should be no connection from collector to emitter while the transistor is unpowered, a short circuit can sometimes develop during a failure.

If any of your junctions read an unusually high (or infinite) resistance in both directions, the junction is probably open circuited. An unusually low resistance (or 0 Ω) in either direction suggests that the junction is short circuited. Any resistance below infinity between the collector and emitter suggests a damaged transistor. In any case, the transistor should be replaced. If you are uncertain how to interpret your readings, test a new transistor or one that you know is good and compare readings.

Checking integrated circuits

There are very few *conclusive* ways to test integrated circuits without resorting to logic analyzers and specialized IC testing equipment. Integrated circuits are so incredibly diverse that there is simply no one universal test that will pinpoint a failure. However, you can make extensive use of IC service charts (sometimes called *service checkout charts*). Integrated circuit service charts show the logic (or voltage) level for each pin of an IC. By checking the actual state of each pin against the chart, you can often identify faulty devices. If you have a schematic (or know the signal patterns) used for an IC, you can often deduce failures. If an expected signal enters an IC, but the expected output signal is absent, it's a good bet that the IC has failed.

Circuit analyzer

If you are a technician or work in a professional computer repair environment, it is faster and more helpful for you to test components while still inserted in the circuit (or check entire circuit arrangements with a single test). Circuit analyzers such as the Model 545 from B+K Precision (Fig. 3-12) use an *impedance signature* technique to examine parts and circuits. An impedance signature is obtained by applying a variable-current sine wave signal source across a component (or an entire circuit) and then displaying the voltage versus current (VI) curve on a CRT or flat-panel LCD. Every component or circuit offers a unique, repeatable VI curve, so any differences at all are easily recognized. The Model 545 offers dual input which allows the suspect component(s) or circuit to be compared against a sample that you know is good.

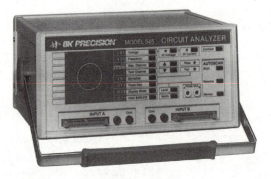

■ **3-12** *A B+K Model 545 Circuit Analyzer.*
(B+K Precision)

The displays illustrated in Fig. 3-13 show the typical output of a circuit analyzer. Test systems such as the Model 545 will test resistors, capacitors, inductors, diodes (and diode bridges), LEDs, zener diodes, bipolar transistors, FETs, optoisolators, and SCRs. It is also interesting to note that the analyzer can test both analog and digital ICs as well as a broad variety of complete circuits. The B+K analyzer offers a serial communication interface that allows images presented on the display to be downloaded to a computer for printing or future reference. Although actual testing procedures are different for every type of system, circuit analyzers are reliable and powerful additions to your professional test bench if your test volume can justify the expense.

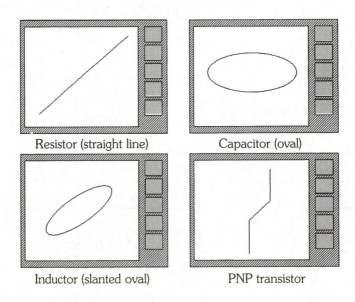

■ **3-13** *Typical curcuit analyzer traces.*

Logic probes

The problem with multimeters is that they do not relate very well at all to the fast-changing or oddly shaped signals found in just about all analog and digital circuits. When a signal changes quickly (e.g., a raster sawtooth wave), a dc voltmeter will only show the *average* signal. A logic probe is little more than an extremely simple voltage sensor, but it can precisely and conveniently detect digital logic levels, clock signals, and digital pulses. Some logic probes can operate at speeds greater than 50 MHz. You may need logic probes when working with TTL monitors.

Logic probes are rather simple-looking devices as shown in Fig. 3-14. Indeed, logic probes are perhaps the simplest and least expensive test instruments that you will ever use, but they provide valuable information when you are troubleshooting digital logic circuitry. Logic probes are usually powered from the circuit under test, and they must be connected into the common (ground) of the circuit being tested to ensure a proper reference level. Attach the probe's power lead to any logic supply voltage source in the circuit. Logic probes are capable of working from a wide range of supply voltages (typically 4 to 18 V dc). Special caution is required when powering the logic probe from a monitor circuit. You must be absolutely certain that you are powering and grounding the probe correctly. **Do not** ground the probe to the monitor's chassis. A chassis is typically "hot" with voltage levels that will destroy the logic probe.

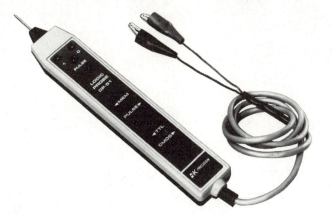

■ **3-14** *A B+K Model DP-51 Logic Analyzer. (B+K Precision)*

Reading the logic probe

Logic probes use a series of LED indicators to display the measured condition: a logic "high" (or 1), a logic "low" (or 0), or a "pulse" (or clock) signal. Many models offer a switch that allows the probe to operate with two common logic families (TTL or CMOS). You may sometimes find TTL and CMOS devices mixed in the same circuit, but one family of logic devices will usually dominate.

To use a logic probe, touch its metal tip to the desired IC or component lead. Be certain that the point you wish to measure is, in fact, a logic point; high-voltage signals in a monitor can easily damage your logic probe. The logic state is interpreted by a few simple gates within the probe and then displayed on the appropriate LED (or combination of LEDs). Table 3-1 illustrates the LED sequences for a B+K Precision logic probe. By comparing the probe's measurements to the information contained in an IC service chart or schematic diagram, you can determine whether or not the suspect IC is operating properly.

■ Table 3-1 Typical logic probe display patterns

Input signal	HI LED	LOW LED	PULSE LED
Logic "1" (TTL or CMOS)	On	Off	Off
Logic "0" (TTL or CMOS)	Off	On	Off
Bad Logic Level or Open Circuit	Off	Off	Off
Square Wave (<200 kHz)	On	On	Blink
Square Wave (>200 kHz)	On/Off	On/Off	Blink
Narrow "High" Pulse	Off	On/Off	Blink
Narrow "Low" Pulse	On/Off	Off	Blink

Oscilloscopes

Oscilloscopes offer a tremendous advantage over multimeters and logic probes. Instead of reading signals in terms of numbers or lighted indicators, an oscilloscope will show voltage versus time on a graphic display. Not only can you observe ac and dc voltages, but oscilloscopes enable you to watch any other unusual signals occur in real time (e.g., raster and analog video signals). When used correctly, an oscilloscope allows you to witness signals and events occurring in terms of microseconds. If you have used an oscilloscope (or seen one being used), then you probably know just how valuable they can be. Oscilloscopes such as the one shown in Fig. 3-15 may appear somewhat overwhelming at first, but many of their operations work the same way regardless of the model in use.

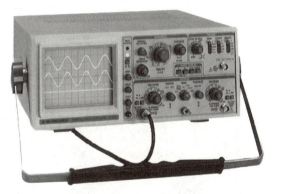

■ **3-15** *A B+K Model 2190A digital oscillo-scope. (B+K Precision)*

Oscilloscope startup procedure

Before you begin taking measurements, a clear, stable trace must be obtained (if not already visible). If a trace is not already visible, make sure that any CRT screen storage modes are turned off and that trace intensity is turned up to at least 50 percent. Set trace triggering to its automatic mode and adjust the horizontal and vertical offset controls to the center of their ranges. Be sure to select an *internal* trigger source from the channel your probe is plugged in to and then adjust the trigger *level* until a stable trace is displayed. Vary your vertical offset if necessary to center the trace in the CRT.

If a trace is not yet visible, use the *beam finder* to reveal the beam's location. A beam finder simply compresses the vertical and horizontal ranges to force a trace onto the display. This gives you a rough idea of the trace's relative position. Once you are finally able to move the trace into position, adjust your focus and intensity controls to obtain a crisp, sharp trace. Keep intensity at a moderately low level to improve display accuracy and preserve the CRT phosphors.

Your oscilloscope should be calibrated to its probe before use. A typical oscilloscope probe is shown in Fig. 3-16. Calibration is a quick and straightforward operation that requires only a low-amplitude, low-frequency square wave. Many models have a built-in "calibration" signal generator (usually a 1-kHz, 300-mV square wave with a duty cycle of 50 percent). Attach your probe to the desired input jack; then place the probe tip across the calibration signal. Adjust your horizontal (TIME/DIV) and vertical (VOLTS/DIV) controls so that one or two complete cycles are clearly shown on the CRT.

Observe the visual characteristics of the test signal as shown in Fig. 3-17. If the square wave's corners are rounded, there may not

■ **3-16** *A B+K Model PR-46 10:1 oscilloscope probe. (B+K Precision)*

be enough probe capacitance (sometimes denoted with the label *Cprobe*). Spiked square wave corners suggest too much capacitance in the probe. Either way, the scope and probe are not matched properly. You must adjust the probe capacitance to establish a good electrical match; otherwise, signal distortion will re-

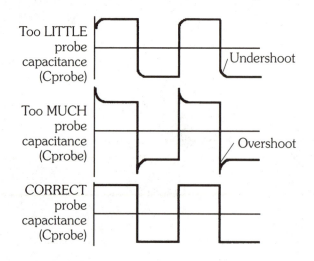

■ **3-17** *A comparison of typical oscilloscope probe calibration signals.*

sult during your measurements. Slowly adjust the variable capacitance of your probe until the corners shown on the calibration signal are as square as possible. If you are not able to achieve a clean square wave, try a different probe.

Monitor cautions (please read this)

NEVER, under any circumstances, should you connect your oscilloscope ground to the monitor's chassis. In actual operation, the monitor chassis is typically isolated from ground and can float at relatively high voltages. Because the oscilloscope ground is actually earth ground, connecting this ground to a hot chassis will result in damage to your oscilloscope (and possible personal injury). Ground your oscilloscope at the power supply ground.

Also, most basic oscilloscopes limit input voltages to 400 V or so (check the specifications for your own particular oscilloscope). Since voltages in the monitor can easily exceed 400 V at certain points, use **extreme** caution when choosing measurement points. Excessive voltage at the input can damage your oscilloscope. It is certainly worthwhile to acquire a set of schematics for the monitor before starting to take measurements.

Measuring voltage

The first step in any voltage measurement is to set your normal trace (called the *baseline*) where you want it. Normally, the baseline is placed along the center of the graticule during start-up, but it can be placed anywhere along the CRT so long as the trace is visible. To establish a baseline, switch your input coupling control to its ground position. Grounding the input disconnects any existing input signal and ensures a zero reading. Adjust the vertical offset control to shift the baseline wherever you want the zero reading to be (usually in the display center). If you have no particular preference, simply center the trace in the CRT.

To measure dc, set your input coupling switch to its DC position; then adjust the VOLTS/DIV control to provide the desired amount of sensitivity. If you are unsure just which sensitivity is appropriate, start with a very low sensitivity (a large VOLTS/DIV setting) and then carefully increase it (reduce the VOLTS/DIV setting) after your input signal is connected. This procedure prevents a trace from simply jumping off the screen when an unknown signal is first applied. If your signal does happen to leave the visible portion of the display, you could reduce the sensitivity (increase the VOLTS/DIV setting) to make the trace visible again.

83

For example, suppose you were measuring a +5 V dc power supply output. If VOLTS/DIV is set to 5 VOLTS/DIV, each major vertical division of the CRT display represents 5 V dc, so your +5 V dc signal should appear one full division above the baseline (5 VOLTS/DIV × 1 DIV = 5 V dc) as shown in Fig. 3-18. At a VOLTS/DIV setting of 2 VOLTS/DIV, the same +5-V dc signal would now appear two and a half divisions above your baseline (2 VOLTS/DIV × 2.5 DIV = 5 V dc). If your input signal were a negative voltage, the trace would appear *below* the baseline, but it would be read the same way.

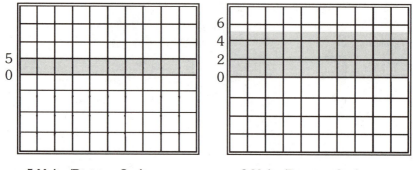

5 Volts/Division Scale   2 Volts/Division Scale

■ **3-18** *Measuring dc voltages with an oscilloscope.*

Alternating current signals can also be read directly from the oscilloscope. Switch your input coupling control to its AC position; then set a baseline just as you would for dc measurements. If you are unsure about how to set the vertical sensitivity, start with a low sensitivity (a large VOLTS/DIV setting) and then slowly increase it (reduce the VOLTS/DIV setting) once your input signal is connected. Keep in mind that ac voltage measurements on an oscilloscope will *not* match ac voltage readings on a multimeter. An oscilloscope displays instantaneous *peak* values for a waveform, whereas ac voltmeters measure in terms of root mean square (RMS) values. To convert a peak voltage reading to RMS, divide the peak reading by 1.414. Another limitation of multimeters is that they can only measure sinusoidal ac signals. Square, triangle, or other unusual waveforms will be interpreted as an average value by a multimeter.

When actually measuring an ac signal, it may be necessary to adjust the oscilloscope's trigger level control to obtain a stable (still) trace. As Fig. 3-19 illustrates, signal voltages can be measured directly from the display. For example, the sinusoidal waveform of Fig. 3-19 varies from −10 to +10 V. If oscilloscope sensitivity were set to 5 VOLTS/DIV, signal peaks would occur two divisions above and two divisions below the baseline. Since the oscilloscope provides peak measurements, an ac voltmeter would show the signal as peak/1.414 (10/1.414) or 7.07 V RMS.

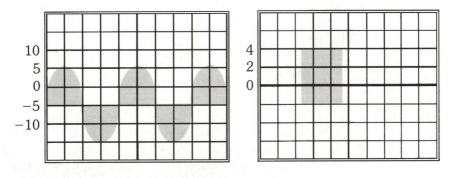

■ **3-19** *Measuring ac voltages with an oscilloscope.*

Measuring time and frequency

An oscilloscope is an ideal tool for measuring critical parameters such as pulse width, duty cycle, and frequency. It is the horizontal sensitivity control (TIME/DIV) that comes into play with time and frequency measurements. Before making any measurements, you must first obtain a clear baseline as you would for voltage measurements. When a baseline is established and a signal is finally connected, adjust the TIME/DIV control to display one or two complete signal cycles.

Typical period measurements are illustrated in Fig. 3-20. With VOLTS/DIV set to 5 ms/DIV, the sinusoidal waveform shown repeats

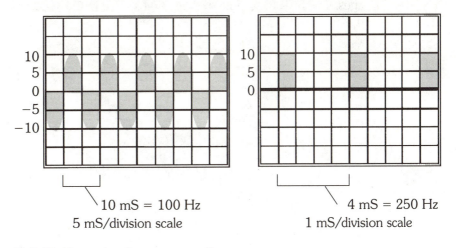

10 mS = 100 Hz
5 mS/division scale

4 mS = 250 Hz
1 mS/division scale

■ **3-20** *Measuring time on an oscilloscope.*

every two divisions. This represents a period of 10 ms (5 ms/DIV × 2 DIV). Since frequency is simply the reciprocal of time, it can be calculated by inverting the time value. A period of 10 ms would represent a frequency of 100 Hz (1/10 ms). This also works for square waves and regularly repeating nonsinusoidal waveforms. The square wave shown in Fig. 3-20 repeats every four divisions. At a TIME/DIV setting of 1 ms/DIV, its period would be 4 ms. This corresponds to a frequency of 250 Hz (1/4 ms).

Instead of measuring the entire period of a pulse cycle, you can also read the time *between* any two points of interest. For the square wave of Fig. 3-20, you could read the pulse width to be 1 ms. You could also read the low portion of the cycle as a duration of 3 ms (added together for a total signal period of 4 ms). A signal's *duty cycle* is simply the ratio of a signal's *on* time to its total period expressed as a percentage. For example, a square wave that is on for 2 ms and off for 2 ms would have a duty cycle of 50% [2 ms/(2 ms + 2 ms) × 100%]. For an on time of 1 ms and an off time of 3 ms, its duty cycle would be 25% [1 ms/(1 ms + 3 ms) × 100%], and so on.

High-voltage probes

Although multimeters and oscilloscopes are handy for measuring potentials up to several hundred volts, they are totally inappropriate instruments for measuring high-voltage at the CRT anode. Under normal operation, anode voltage can be anywhere from 15 to 30 kV or more depending on the size and vintage of the individual monitor. A high-voltage probe is a specially designed test instrument that can safely and effectively measure anode voltages. The B+K Precision Model HV-44 shown in Fig. 3-21 can handle up to 40 kV. Since excessive high voltage can present an

■ **3-21** *A B+K Model HV-44 high-voltage probe-meter. (B+K Precision)*

X-radiation danger, it is important that you check the anode voltage as the very first step in every monitor repair. If you do not have a high-voltage probe, it is a worthwhile investment for any serious monitor worker.

Using a high-voltage probe is every bit as easy as using a multimeter. Connect the probe's ground clip to power supply ground and then gently slip the probe tip beneath the rubber anode cover until you reach the anode's metal conductors. Take your high-voltage reading directly from the meter. The B+K HV-44 provides you with two interchangeable metal probe tips.

Probe safety

Even though high-voltage probes are designed to operate safely at very high voltages, there are still a number of **important** safety considerations to keep in mind. First, inspect the high-voltage probe very carefully before using it. Its insulating properties may be compromised if there are any chips or cracks in the plastic housing. The probe surface must also be completely clean and dry. Make sure that the metal probe tip is inserted properly and completely, but do not use hand tools to tighten the tip. Scratches in the tip can act as discharge points for high voltage. Use only one hand to hold the probe and keep your other hand behind your back or in a pocket. Serious shocks can still occur if the probe's insulation should breach while you are holding onto a ground conductor with the other hand. Finally, commercial high-voltage probes are designed for measuring positive voltages. Connecting a negative or alternating voltage can damage the probe.

CRT testers-restorers

In spite of their size and complexity, CRTs are basically nothing more than big vacuum tubes. Like all vacuum tubes, they gradually degrade and eventually fail. Degrading CRTs can reduce picture quality, brightness, and sharpness. When supply voltages and electrical signal levels measure properly, the CRT becomes suspect. You can simply replace the CRT when troubleshooting symptomatically. When you have to deal with a large volume of monitors (especially older monitors), it is often worth investing in a full-featured CRT tester-restorer. The B+K Precision Model 490 (Fig. 3-22) is one such piece of test equipment.

The Model 490 is designed for use with most monochrome and color CRTs. In actual use, the 490 tests for emission, leakage,

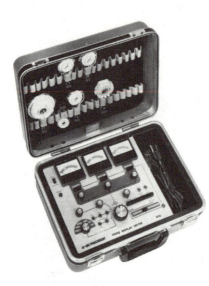

tracking, life, and focus. These five tests provide a comprehensive picture of the CRT's overall condition. *Emission* tests the CRT's electron gun(s) to see that each one can produce adequate brightness on the picture tube. *Leakage* tests check the CRT's cathode(s) for partial short circuits. Since cathodes provide the source for an electron beam, leakage can reduce the brightness of an image or result in too little of a primary color. A *life expectancy* test is performed when the results of an emission test are marginal. *Tracking* checks that each of the three electron guns in a color CRT exhibit roughly equal performance as their excitation signals vary. If the guns do not perform equally, variations in monitor power may cause color fluctuations in the screen image. The *focus* test checks the integrity of a CRT's focus grid(s).

The Model 490 can also perform a limited amount of CRT rejuvenation by clearing shorts, cleaning and balancing electron guns, and restoring cathodes. While rejuvenation will not make a CRT as "good as new," it can restore a monitor's operation for a while longer and save the customer the expense of a new CRT installation (often approaching the cost of a new monitor). Electron gun assemblies are extremely small and tightly packaged devices. Over time, coating materials can become loose and short with adjacent elements. The *remove shorts* feature uses a high energy discharge from a capacitor to "zap" any shorts that may be causing excessive leakage. The *rejuvenate cathode* function can be used after a low or marginal emission test. Rejuvenation restores a cathode's operation by applying a large positive voltage between the cathode and the controlling grid.

Monitor testers-analyzers

Since we often judge the performance of monitors based on what we see on the screen, image quality can easily become a subjective matter. Monitor testers provide a higher class of test equipment that allows you to test and align the monitor using standard visual patterns. Some testers also incorporate meter functions that enable you to actually perform circuit measurements while the equipment is connected.

Network Technologies, Inc. (NTI)

The Network Technologies MONTEST-AD16 (Fig. 3-23) is one such monitor tester. The AD16 is designed to offer four major test patterns at 16 horizontal scanning frequencies ranging from 15.7 to 64.0 kHz. The AD16 supports VGA, MAC II, MCA, CGA, EGA, PGA, Hewlett-Packard, MicroVax, and Apollo monitors. Both TTL and analog video signals are provided. A *crosshatch* (or Xhatch) pattern is used to test convergence and focus. The *raster* pattern tests color purity. An array of *color bars* tests color balance between various colors as well as overall screen intensity settings. A *window* pattern helps discern the condition of a monitor's high-voltage system. Network Technologies offers several different models of MONTEST monitor testers. At about the price of a good oscilloscope, NTI's line of testers is a handy addition to the professional test bench.

■ **3-23** *An NTI Montest-AD16 video test instrument. (Courtesy of NTI)*

Sencore

Sencore is probably one of the most recognized and respected names in high-end consumer electronics diagnostic tools. Their CM2125 (Fig. 3-24) represents the state of the art in monitor repair test instruments. Like other monitor testers, the CM2125 provides a rich variety of video patterns for testing and alignment. It also incorporates digital meter instrumentation that allows technicians to test the monitor circuit at points from the video input to the high-voltage circuit. Video, vertical sync, vertical drive, horizontal sync, and horizontal drive signals can be injected into the monitor being tested to check each section of the circuit. A specialized "ringer" generator allows you to check flyback transformers and deflection yokes. If you are repairing computer monitors for a living, the CM2125 is largely considered the ultimate component-level troubleshooting tool. Make no mistake, the CM2125 is not for the timid. Its price tag alone limits it to the most serious professional repair facilities.

■ **3-24** *A Sencore CM2125 monitor analyzer. (Sencore)*

Computer-aided troubleshooting

Monitor troubleshooting is certainly nothing new, and there are several organizations that have developed to share their troubleshooting expertise in the form of knowledge-based programs or troubleshooting indexes. While these computer-aided troubleshooting products tend to be rather expensive for individuals, professional technicians and independent repair shops can reap some real rewards from the accumulated experience that is available today.

Resolve database

Anatek Corporation has developed the *Resolve Monitor Repair Database*. By selecting the manufacturer and model of the defective monitor, a list of common symptoms is displayed. You can then select the appropriate symptom and review the suggested corrective action. In most cases, repair suggestions point directly to the suspect component(s). If more than one component is suspect, they are listed in order of probability. Because Resolve is a database (using the Paradox run-time engine), you can also enter your own symptoms and solutions as well as edit and expand existing symptoms. The commercial package contains over 1900 repair procedures along with sources for components, a manufacturer's reference, FCC identification numbers, and telephone numbers.

You can try the Resolve database by downloading the shareware version from TechNet BBS at 508-366-7683. You will need three files: RESOLVE1.ZIP, RESOLVE2.ZIP, and RESREAD.ME. The shareware version offers over 700 procedures for 81 different monitors. To operate the shareware version, you need an i286-class PC or later with DOS, 2 Mbytes of RAM (8 Mbytes for optimum performance), and 5 Mbytes of hard drive space. For more information on the Resolve Monitor Repair Database, you can contact:

Anatek Corp
P.O. Box 1200
Amherst, NH 03031-1200
Tel: 603-673-4342
E-mail: info@anatekcorp.com
WWW: http://www.dlspubs.com/

IMPACT

IVS offers the *IMPACT* (Instructional Monitor Problem Analysis and Correction Toolkit) package. IMPACT is fundamentally a Windows-based database focusing on a collection of over 200 fully described monitor troubleshooting procedures and specific repairs. IMPACT goes beyond monitor-specific solutions to provide basic repair procedures, in-depth explanations of monitor operations, and component-level testing. The program is expandable, so you can easily add your own notes and repair procedures to the database. There is no shareware version available, but for more information and pricing, you can contact:

IVS
13311 Stark Rd.
Livonia, MI 48150
Tel: 313-261-8801

4

Video adapters

THE MONITOR ITSELF IS MERELY AN OUTPUT DEVICE (A peripheral) that translates synchronized analog or transistor-transistor logic (TTL) video signals into a visual image. Of course, a monitor *alone* is not good for very much, except perhaps as a conversation piece or a room heater. The first logical question is: Where does the video signal come from? All video signals displayed on a monitor are produced by a *video adapter* circuit (Fig. 4-1). The term *adapter* is used because the PC is "adapted" to the particular monitor through this circuit. In most cases, the video adapter is an expansion board that plugs into the PC's expansion bus connectors. It is the video adapter that converts raw data from the PC into image data which is stored in the adapter's *video memory*. The exact amount of memory available depends on the particular adapter and the video modes that the adapter is designed to support. Simple adapters offer as little as 256 kbytes, whereas the latest adapters provide 2 Mbytes or more. The video adapter then translates the contents of video memory into corresponding video signals that drive a monitor.

While the actual operations of a video adapter are certainly more involved than just described, you can begin to appreciate the critical role that the video adapter plays in a PC. If a video adapter fails, the monitor will display gibberish (or nothing at all). To complicate matters even further, many current software applications require small device drivers (called *video drivers*). A video driver is a rather small program that allows an application to access a video adapter's high-resolution or high-color video modes (usually for SVGA operation). During troubleshooting, it will be necessary for you to isolate display problems to either the monitor, the video adapter, or driver software before a solution can be found.

Understanding conventional video adapters

The conventional frame buffer is the oldest and most well-established type of video adapter. The term *frame buffer* refers to

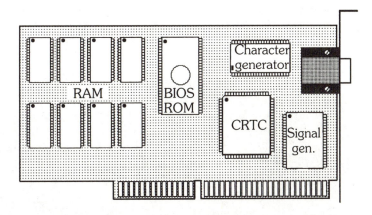

■ 4-1 *A typical video adapter board.*

the adapter's operation: Image data are loaded and stored in video memory one "frame" at a time. Frame buffer architecture (as shown in Fig. 4-2) has changed very little since PCs first started displaying text and graphics. The heart of the frame buffer video adapter is the highly integrated display controller IC (sometimes called a CRTC or *cathode-ray tube controller*). The CRTC generates control signals and supervises adapter operation. It is the CRTC that reads *video RAM* (or VRAM) contents and passes those contents along for fur-

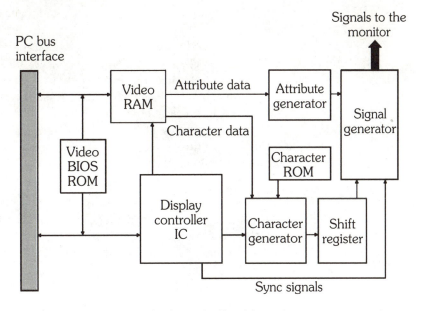

■ 4-2 *Block diagram of a frame buffer video adapter.*

ther processing. Many new video boards use specially designed IC groups (called *chipsets*) which are intended to work together. Chipsets provide fast, efficient video performance while minimizing the amount of circuitry needed on a video adapter.

Text versus graphics

Video RAM also plays a vital role because it is RAM that holds the image data to be displayed. The video adapter can operate in two modes: text and graphic. Let's start by understanding the *text* mode. ASCII characters are stored in video RAM. A character ROM, character generator, and shift register produce the pixel patterns that form ASCII screen characters. The character ROM holds a pixel pattern for every possible ASCII character (including letters, numbers, and punctuation). The character generator converts ROM data into a sequence of pixel bits and transfers them to a shift register. The shift register produces a bit stream. At the same time, an attribute decoder determines whether the defined ASCII character is to be displayed as blinking, inverted, high intensity, standard text, or a text color (for color monitors). The signal generator is responsible for turning the ASCII serial bit stream from the shift register into the video and synchronization signals that actually drive the monitor. The signal generator may produce either analog or TTL video signals.

In the *graphic* mode, video RAM locations already contain the color-gray scale information for each screen pixel rather than ASCII characters, so the character ROM and character generating circuitry used in text mode are bypassed. For example, monochrome graphics use a single bit per pixel, 16 color graphics use 4 bits per pixel, 256 color graphics use 8 bits per pixel, and so on. Pixel data taken from VRAM by the CRTC are passed through the character generator without any changes. Data are then sent directly to the shift register and on to the signal generator. It is the signal generator that produces analog or TTL video signals along with sync signals as dictated by the CRTC.

ROM BIOS (video BIOS)

There is one part of the classical video adapter which has not been mentioned yet: the ROM BIOS (also known as the video BIOS). The display controller requires substantial instruction changes when it is switched from text mode to any one of its available graphics modes. Since the instructions required to reconfigure and direct the CRTC depend on its particular design (and the chipsets in use), it is impossible to rely on the software application or the PC's BIOS to provide the required software. As a result, all video adapters from EGA on use local BIOS ROM to hold the firmware needed by the particular display controller. Video adapters are configured to position the video BIOS within 128 kbytes of space from C0000h to DFFFFh. This space is rarely used by the video BIOS entirely because it is reserved for a variety of devices with expansion ROMs such as hard drive controllers and video adapters. Motherboard BIOS works in conjunction with the video BIOS.

Video memory versus video mode

Graphics demand video memory. This may be an obvious statement, but determining the *amount* of memory required to hold a screen image is not so intuitive. There must be enough video memory to represent the color of each available pixel. Consider the standard VGA screen mode at $640 \times 480 \times 16$. There are 307,200 (640×480) pixels, but each pixel can be any of 16 colors. Four bits are needed to define 1 of 16 colors, so 1,228,800 ($307,200 \times 4$) bits are needed. Since it is more convenient to express memory in bytes, divide the number of bits by 8. When you do the math, you see that 153,600 bytes (1,228,800 / 8), or 153.6 kbytes, are needed for a single VGA frame. That's not a lot of memory for a video board.

Consider what happens when you need to display a "true-color" image (16M colors) at 640×480. You still have the same 307,200 (640×480) pixels to deal with, but there are now 16,777,216 possible colors. It takes 3 bytes (24 bits) to define a number between 0 and 16,777,215, so you would need 921.6 kbytes ($307,200 \times 3$) of video memory to hold one complete screen image in that video mode. Table 4-1 illustrates the memory requirements for typical video modes and shows you the minimum amount of memory needed on the video board for each mode.

This relationship between screen modes and video memory becomes very important when choosing a video board or choosing a screen mode to operate your video board in. For example, a video board with 512 kbytes of video memory will not support more than 256 colors in 640×480 mode or 800×600 mode. It will also not support more than 16 colors in 1024×768 mode. On the other hand, a video board with 4 Mbytes of video memory will support "true-color" images in resolutions up to 1280×1024.

Reviewing video display hardware

The early days of PC development left users with a simple choice between monochrome or color graphics (all video adapters sup-

Resolution	Colors	Memory requirements	
		1 page	Board
640 × 480	16	153.6 kbytes	(256 kbytes)
640 × 480	256	307.2 kbytes	(512 kbytes)
640 × 480	65,536	614.4 kbytes	(1 Mbyte)
640 × 480	16,777,216	921.6 kbytes	(1 Mbyte)
800 × 600	16	240.0 kbytes	(256 kbytes)
800 × 600	256	480.0 kbytes	(512 kbytes)
800 × 600	65,536	960.0 kbytes	(1 Mbyte)
800 × 600	16,777,216	1440.0 kbytes	(2 Mbytes)
1024 × 768	16	393.2 kbytes	(512 kbytes)
1024 × 768	256	786.4 kbytes	(1 Mbyte)
1024 × 768	65,536	1572.8 kbytes	(2 Mbytes)
1024 × 768	16,777,216	2359.3 kbytes	(4 Mbytes)
1280 × 1024	16	655.4 kbytes	(1 Mbyte)
1280 × 1024	256	1310.7 kbytes	(2 Mbytes)
1280 × 1024	65,536	2621.4 kbytes	(4 Mbytes)
1280 × 1024	16,777,216	3932.1 kbytes	(4 Mbytes)

port text modes). In the years that followed, however, the proliferation of video adapters has brought an array of video modes and standards that you should be familiar with before upgrading a PC or attempting to troubleshoot a video system. This part of the chapter explains each of the video standards that have been developed in the last 15 years and shows you the video modes that each standard offers.

MDA (monochrome display adapter, 1981)

The monochrome display adapter is the oldest video adapter available for the PC. Text is available in 80 column × 25 row format using 9 × 14 pixel characters as shown in Table 4-2. Being a text-only system, MDA offered no graphics capability, but it achieved popularity because of its relatively low cost, good text display quality, and integrated printer port. Figure 4-3 shows the video connector pinout for an MDA board. The nine-pin monitor connection uses four active TTL signals: intensity, video, horizontal, and vertical. *Video* and *intensity* signals provide the on/off and high/low intensity information for each pixel. The *horizontal* and *vertical* signals control the monitor's synchronization. Monochrome display adapter boards have long been obsolete, and the probability of your encountering one is remote at best.

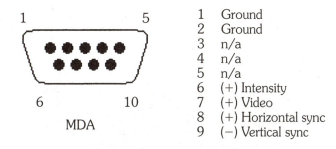

1	Ground
2	Ground
3	n/a
4	n/a
5	n/a
6	(+) Intensity
7	(+) Video
8	(+) Horizontal sync
9	(−) Vertical sync

MDA

■ **4-3** *Pinout of an MDA video connector.*

CGA (color graphics adapter, 1981)

The color graphics adapter was the first to offer color text and graphics modes for the PC. A 160 × 200 low-resolution mode offered 16 colors, but such low resolution received very little attention. A 320 × 200 medium-resolution graphics mode allowed finer graphic detail, but with only 4 colors. The highest resolution mode provided 640 × 200 at 2 colors (usually black and one other color). The relationship between resolution and colors is important since a CGA *frame* requires 16 kbytes of video RAM. A resolution of 640 × 200 results in 128,000 pixels. With 8 bits able to represent 8 pixels, 16,000 (128,000/8) bytes are adequate. A resolution of 320 × 200 results in 64,000 pixels, but with 2 bits needed to represent 1 pixel (4 pixels/byte), 16,000 (64,000/4) bytes are still enough. You can see that video RAM is directly related to video capacity. Since there is typically much more video RAM available than is needed for an image, video boards support multiple video *pages*.

Figure 4-4 shows the pinout for a typical CGA video connector. As with the earlier MDA design, CGA video signals reserve pins 1 and 2 as ground lines; the horizontal sync signal is produced on pin 8, and the vertical sync signal is produced on pin 9. CGA is strictly a digital display system with TTL signals used on the red (3), green (4), blue (5), and intensity (6) lines.

You may also encounter CGA monitors and video boards fitted with a phono jack (i.e., two-wire) connection rather than the standard nine-pin connector, and this is a *composite CGA connector*. A composite connector places all three color signals (along with sync information) on the same wire in an analog format. Composite configurations simplify cabling, but additional circuitry in the monitor is required to separate the color and sync signals. The process of separation often results in reduced signal quality. Given the age of

97

■ Table 4-2 Comparison of video modes and standards

Standard	Resolution(s)	Colors	Mode	Text Fmt.	Vert. scan	Horiz. scan	BIOS Mode
MDA	720 × 350	n/a	Text	80 × 25	50 Hz	18.432 kHz	07h
CGA	320 × 200	16	Text	40 × 25	60 Hz	15.750 kHz	00h/01h
	640 × 200	16	Text	80 × 25	60 Hz	15.750 kHz	02h/03h
	160 × 200	16	Graphics	n/a	60 Hz	15.750 kHz	n/a
	320 × 200	4	Graphics	40 × 25	60 Hz	15.750 kHz	04h/05h
	640 × 200	2	Graphics	80 × 25	60 Hz	15.750 kHz	06h
EGA	320 × 350	16	Text	40 × 25	60 Hz	21.850 kHz	00h/01h
	640 × 350	16	Text	80 × 25	60 Hz	21.850 kHz	02h/03h
	720 × 350	4	Text	80 × 25	50 Hz	18.432 kHz	07h
	320 × 200	16	Graphics	40 × 25	60 Hz	15.750 kHz	0Dh
	620 × 200	16	Graphics	80 × 25	60 Hz	15.750 kHz	0Eh
	640 × 350	4	Graphics	80 × 25	50 Hz	18.432 kHz	0Fh
	640 × 350	16	Graphics	80 × 25	60 Hz	21.850 kHz	10h
PGA	320 × 200	16	Text	40 × 25	60 Hz	15.750 kHz	00h/01h
	640 × 200	16	Text	80 × 25	60 Hz	15.750 kHz	02h/03h
	320 × 200	4	Graphics	40 × 25	60 Hz	15.750 kHz	04h/05h
	640 × 200	2	Graphics	80 × 25	60 Hz	15.750 kHz	06h
	640 × 480	256	Graphics	n/a	60 Hz	30.480 kHz	n/a
MCGA	320 × 400	16	Text	40 × 25	70 Hz	31.500 kHz	00h/01h
	640 × 400	16	Text	80 × 25	70 Hz	31.500 kHz	02h/03h
	320 × 200	4	Graphics	40 × 25	70 Hz	31.500 kHz	04h/05h
	640 × 200	2	Graphics	80 × 25	70 Hz	31.500 kHz	06h
	640 × 480	2	Graphics	80 × 30	60 Hz	31.500 kHz	11h
	320 × 200	256	Graphics	40 × 25	70 Hz	31.500 kHz	13h
VGA	360 × 400	16	Text	40 × 25	70 Hz	31.500 kHz	00h/01h
	720 × 400	16	Text	80 × 25	70 Hz	31.500 kHz	02h/03h
	320 × 200	4	Graphics	40 × 25	70 Hz	31.500 kHz	04h/05h

Adapter	Resolution	Colors	Type	Characters	Vertical	Horizontal	Mode
	640 × 200	2	Graphics	80 × 25	70 Hz	31.500 kHz	06h
	720 × 400	16	Text	80 × 25	70 Hz	31.500 kHz	07h
	320 × 200	16	Graphics	40 × 25	70 Hz	31.500 kHz	0Dh
	640 × 200	16	Graphics	80 × 25	70 Hz	31.500 kHz	0Eh
	640 × 350	4	Graphics	80 × 25	70 Hz	31.500 kHz	0Fh
	640 × 350	16	Graphics	80 × 25	70 Hz	31.500 kHz	10h
	640 × 480	2	Graphics	80 × 30	60 Hz	31.500 kHz	11h
	640 × 480	16	Graphics	80 × 30	60 Hz	31.500 kHz	12h
	320 × 200	256	Graphics	40 × 25	70 Hz	31.500 kHz	13h
8514	1024 × 768	256	Graphics	85 × 38	43.48 Hz	35.520 kHz	n/a
	640 × 480	256	Graphics	80 × 34	60 Hz	31.500 kHz	n/a
	1024 × 768	256	Graphics	146 × 51	43.48 Hz	35.520 kHz	n/a
XGA	360 × 400	16	Text	40 × 25	70 Hz	31.500 kHz	00h/01h
	720 × 400	16	Text	80 × 25	70 Hz	31.500 kHz	02h/03h
	320 × 200	4	Graphics	40 × 25	70 Hz	31.500 kHz	04h/05h
	640 × 200	2	Graphics	80 × 25	70 Hz	31.500 kHz	06h
	720 × 400	16	Text	80 × 25	70 Hz	31.500 kHz	07h
	320 × 200	16	Graphics	40 × 25	70 Hz	31.500 kHz	0Dh
	640 × 200	16	Graphics	80 × 25	70 Hz	31.500 kHz	0Eh
	640 × 350	4	Graphics	80 × 25	70 Hz	31.500 kHz	0Fh
	640 × 350	16	Graphics	80 × 25	70 Hz	31.500 kHz	10h
	640 × 480	2	Graphics	80 × 30	60 Hz	31.500 kHz	11h
	640 × 480	16	Graphics	80 × 30	60 Hz	31.500 kHz	12h
	320 × 200	256	Graphics	40 × 25	70 Hz	31.500 kHz	13h
	1056 × 400	16	Text	132 × 25	70 Hz	31.500 kHz	14h
	1056 × 400	16	Text	132 × 43	70 Hz	31.500 kHz	14h
	1056 × 400	16	Text	132 × 56	70 Hz	31.500 kHz	14h
	1056 × 400	16	Text	132 × 60	70 Hz	31.500 kHz	14h
	1024 × 768	256	Graphics	85 × 38	43.48 Hz	35.520 kHz	14h
	640 × 480	65536	Graphics	80 × 34	60 Hz	31.500 kHz	n/a
	1024 × 768	256	Graphics	128 × 54	43.48 Hz	35.520 kHz	n/a
	1024 × 768	256	Graphics	146 × 51	43.48 Hz	35.520 kHz	n/a

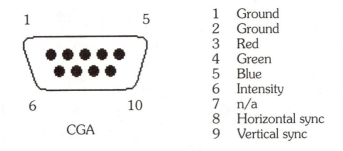

```
1    5


6         10
    CGA
```

1   Ground
2   Ground
3   Red
4   Green
5   Blue
6   Intensity
7   n/a
8   Horizontal sync
9   Vertical sync

■ **4-4** *Pinout of a CGA video connector.*

CGA products, it is unlikely that you should ever encounter this setup, but you should at least be able to recognize it on sight.

EGA (enhanced graphics adapter, 1984)

It was not long before the limitations of CGA became painfully apparent. The demand for higher resolutions and color depths drove designers to introduce the next generation of video adapter known as the enhanced graphics adapter. One of the unique appeals of EGA was its backward compatibility: An EGA board would emulate CGA and MDA modes on the proper monitor as well as its native resolutions and color depths when using an EGA monitor. Enhanced graphics adapter (EGA) is known for its $320 \times 200 \times 16$, $640 \times 200 \times 16$, and $640 \times 350 \times 16$ video modes. More memory is needed for EGA, and 128 kbytes are common for EGA boards (although many boards could be expanded to 256 kbytes).

The EGA connector pinout is illustrated in Fig. 4-5. Transistor-transistor logic (TTL) signals are used to provide primary red (3), primary green (4), and primary blue (5) color signals. By adding a set of secondary color signals (or color *intensity* signals) such as

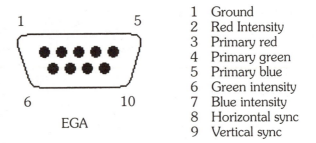

```
1    5


6         10
    EGA
```

1   Ground
2   Red Intensity
3   Primary red
4   Primary green
5   Primary blue
6   Green intensity
7   Blue intensity
8   Horizontal sync
9   Vertical sync

■ **4-5** *Pinout of an EGA video connector.*

red intensity (2), green intensity (6), and blue intensity (7), the total of six color control signals allow the EGA to produce up to 64 possible colors. Although 64 colors are possible, only 16 of those colors are available in the palette at any one time. Pin 8 carries the horizontal sync signal, pin 9 carries the vertical sync signal, and pin 1 remains ground.

PGA (professional graphics adapter, 1984)

The professional graphics adapter was also introduced in 1984. This system offered a then-revolutionary display capability of 640 × 480 × 256. Three-dimensional rotation and graphic clipping was included as a hardware function, and the adapter could update the display at 60 frames per second. The PGA was incredibly expensive and beyond reach of all but the most serious business user. In actual operation, a PGA system required two or three expansion boards, so it also represented a serious commitment of limited system space. Ultimately, PGA failed to capture any significant market acceptance. It is unlikely that you will ever encounter a PGA board because most that ever saw service in PCs have since been upgraded.

MCGA (multicolor graphics array, 1987)

The multicolor graphics array had originally been integrated into the motherboard of IBM's PS/2-25 and PS/2-30. As Table 4-2 shows, MCGA supports all of the CGA video modes and also offers several new video modes including a 320 × 200 × 256 mode that has become a preferred mode for game software. Multicolor graphics array (MCGA) was one of the first graphic systems to use analog color signals rather than TTL signals. Analog signals were necessary to allow MCGA to produce its 256 colors using only three primary color lines.

IBM also took the opportunity to employ a new high-density 15-pin connector as shown in Fig. 4-6. One of the striking differences between the "analog" connector and older TTL connectors is the use of individual ground lines for each color. Careful grounding is vital since any signal noise on the analog lines will result in color anomalies. If you inspect a video cable closely, you will find that one or both ends are terminated with a square metal box which actually contains a noise filter. It is important to realize that, although the MCGA could emulate CGA modes, older TTL monitors were no longer compatible.

There were a number of notable technical improvements that went into the PS/2, but none of them could save the PS/2 line (which is now discontinued). However, the MCGA ushered in a new age of dis-

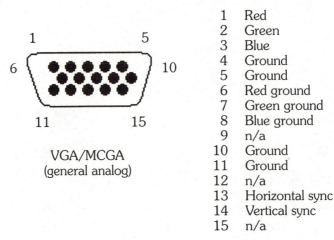

1	Red
2	Green
3	Blue
4	Ground
5	Ground
6	Red ground
7	Green ground
8	Blue ground
9	n/a
10	Ground
11	Ground
12	n/a
13	Horizontal sync
14	Vertical sync
15	n/a

■ **4-6** *Pinout of a VGA/MCGA video connector.*

play technology, and virtually all subsequent video adapters now use the 15-pin analog format shown in Fig. 4-6. While MCGA adapters are also obsolete, the standard lives on in MCGA's cousin, VGA.

VGA (video graphics array, 1987)

The video graphics array was introduced along with MCGA and implemented in other members of the PS/2 line. The line between MCGA and VGA has always been a bit fuzzy because both were introduced simultaneously (both using the same 15-pin video connector), and VGA can handle every mode that MCGA could. For all practical purposes, we can say that MCGA is a *subset* of VGA.

It is VGA that provides the familiar 640 × 480 × 16 screen mode which has become the baseline for Microsoft Windows displays. The use of analog color signals allows VGA systems to produce a palette of 16 colors from 262,144 possible colors. As you see in Table 4-2, VGA also provides backward compatibility for all older screen modes. Although the PS/2 line has been discontinued, the flexibility and backward compatibility of VGA proved so successful that VGA adapters were soon developed for the PC. Video graphics array (VGA) support is now considered to be "standard equipment" for all new PCs sold today, but SVGA boards are rapidly approaching the price of VGA boards, and most SVGA adapters offer full VGA support.

8514 (1987)

The 8514/A video adapter is a high-resolution system also developed for the PS/2. In addition to full support for MDA, CGA, EGA, and VGA modes, the 8514/A can display 256 colors at 640×480 and 1024×768 (interlaced) resolutions. Unfortunately, the 8514/A was a standard ahead of its time. The lack of available software and the demise of the PS/2 line doomed the 8514/A to extinction before it could become an accepted standard. Today, the XGA is rapidly becoming the PC standard for high-resolution–high-color display systems.

SVGA (super video graphics array, 1990)

Ever since VGA became the de facto standard for PC graphics, there has been a strong demand from PC users to move beyond the $640 \times 480 \times 16$ limit imposed by "conventional" VGA to provide higher resolutions and color depths. By 1990, a new generation of extended or *super* VGA (SVGA) adapters had moved into the PC market. Unfortunately, there is no generally accepted standard on which to develop an SVGA board (this is why there is no SVGA reference in Table 4-2). Each manufacturer makes an SVGA board which supports a variety of different, and not necessarily compatible, video modes. For example, one manufacturer may produce a SVGA board capable of $1024 \times 768 \times 65536$, and another may produce a board that only reaches $640 \times 480 \times 16M$ (more than 16 million colors).

This mixing and matching of resolutions and color depths have resulted in a very fractured market; no two SVGA boards are necessarily capable of the same things. This proliferation of video hardware also makes it impossible for applications software to take advantage of *super* video modes without supplemental software called *video drivers*. Video drivers are device drivers (loaded before an application program is started) that allow the particular program to work with the SVGA board hardware. Video drivers are typically developed by the board manufacturer and shipped on a floppy disk with the board. Windows takes particular advantage of video drivers since the Windows interface allows *all* Windows applications to use the same graphics system rather than having to write a driver for every application as DOS drivers must do. Using an incorrect, obsolete, or corrupted video driver can be a serious source of problems for SVGA installations. The one common attribute of SVGA boards is that *most* offer full support for conventional VGA, which requires no video drivers. There are only a handful of SVGA board manufacturers that abandon conventional VGA support.

Today, many SVGA boards offer terrific video performance, a wide selection of modes, and prices that rival high-end VGA adapters. If it were not for the lack of standardization in SVGA adapters, VGA would likely be considered obsolete already. The *Video Electronics Standards Association* (VESA) has started the push for SVGA standards by proposing the VESA BIOS Extension, a universal video driver. The extension would provide a uniform set of functions that allow programmers to detect a card's capabilities and use the optimum adapter configuration regardless of how the particular board's hardware is designed. Many of the quality SVGA boards in production today support the VESA BIOS extensions, and it is worthwhile to recommend boards that support VESA SVGA. Some SVGA boards even incorporate the extensions into the video BIOS ROM which saves the RAM space that would otherwise be needed by a VESA video driver.

XGA (1990)

The XGA and XGA/2 are 32-bit high-performance video adapters developed by IBM to support microchannel based PCs. XGA design with microchannel architecture allows the adapter to take control of the system for rapid data transfers. The XGA standards are shown in Table 4-2. You see that MDA, CGA, EGA, and VGA modes are all supported for backward compatibility. In addition, several color depths are available at 1024×768 resolution, and a photorealistic 65,536 colors are available at 640×480 resolution. To improve performance even further, fast video RAM and a graphics coprocessor are added to the XGA design.

For the time being, XGA is limited to high-performance applications in microchannel systems. The migration to Industry Standard Architecture (ISA) based PCs has been slow because the ISA bus is limited to 16 bits and does not support *bus-mastering* as microchannel busses do. For PCs, SVGA adapters will likely provide extended screen modes as they continue to grow in sophistication as graphics accelerators.

Graphics accelerators and speed factors

When screen resolutions approach 640×480 and beyond, the data needed to form a single screen image can be substantial. Consider a single $640 \times 480 \times 256$ image. There are 307,200 (640×480) pixels. Since there are 256 colors, 8 bits are needed to define the color for each pixel. This means 307,200 bytes are needed for every frame. When the frame must be updated 10 times per second, 3,072,000 ($307,200 \times 10$) bytes per second (3.072 Mbytes/s) must be moved across the ISA bus. If a 65,536 color mode is being used, 2 bytes are needed for each pixel, so 614,400 ($307,200 \times 2$) bytes are needed for a frame. At 10 frames per second, 6,144,000 ($614,400 \times 10$) bytes per second (6.144 Mbytes/s) must be moved across the bus. This is just for video information and does not consider the needs of system overhead operations such as memory refresh, keyboard and mouse handling, drive access, and other data-intensive system operations. When such volumes of information must be moved across an ISA bus limited at 8.33 MHz, you can see how a serious data transfer bottleneck develops. This results in painfully slow screen refreshes—especially under Windows, which requires frequent refreshes.

Video designers seek to overcome the limitations of conventional video adapters by incorporating processing power onto the video board itself rather than relying on the system CPU for graphic processing. By offloading work from the system CPU and assigning the graphics processing to local processing components, graphics performance can be improved by a factor of 3 or more. There are several means of acceleration depending on the sophistication of the board (Fig. 4-7). *Fixed-function acceleration* relieves load on the system CPU by providing adapter support for a limited number of specific functions such as BitBlt or line draws. Fixed-function accelerators were an improvement over frame buffers,

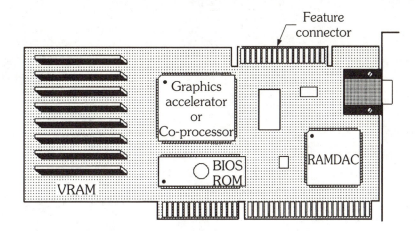

■ **4-7** *A typical video accelerator board.*

but they do not offer the performance of more sophisticated accelerators. A *graphics accelerator* uses an application-specific IC (ASIC) which intercepts graphics tasks and processes them without the intervention of the system CPU. Graphics accelerators are perhaps the most cost-effective type of accelerator. *Graphics coprocessors* are the most sophisticated type of accelerator. The coprocessor acts as a CPU that is dedicated to handling image data. Current graphics coprocessors such as the TMS34010 and TMS34020 represent the Texas Instruments Graphical Architecture (TIGA), which is broadly used for high-end accelerators. Unfortunately, not all graphics coprocessors provide increased performance to warrant the higher cost.

Figure 4-8 shows the block diagram for a typical graphics accelerator. The core of the accelerator is the graphics IC (or chipset). The graphics IC connects directly with the PC expansion bus. Graphics commands and data are translated into pixel data, which are stored in video RAM. High-performance video memory offers a second data bus that is routed directly to the board's random access memory digital-to-analog converter (RAMDAC). The graphics IC directs RAMDAC operation and ensures that VRAM data are available. The RAMDAC then translates video data into red, green, and blue analog signals along with horizontal and vertical synchronization signals. Output signals generated by the RAMDAC drive the monitor.

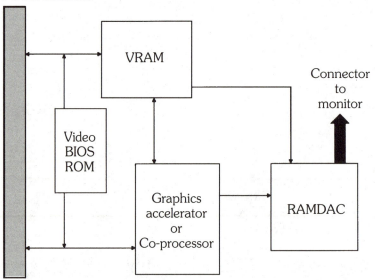

■ **4-8** *Block diagram of a video accelerator board.*

Video speed factors

There is no one element that defines the performance of an accelerator board. Overall performance is actually a combination of five major factors: the video accelerator IC, the video RAM, the video BIOS/drivers, the RAMDAC, and the bus architecture. By understanding how each of these factors relate to performance, you can make the best recommendations for system upgrades or replacement boards. The companion software available for this book provides a utility that checks the specifications of your particular video adapter.

Accelerator ICs and chipsets

Of course, the video accelerator IC itself (or the graphics chipset being used) is at the core of the accelerator board. The type of IC (fixed-function, graphics accelerator, or graphics co-processor) loosely defines the board's capabilities. All other factors being equal, a board with a graphics accelerator will certainly perform better than a fixed-function accelerator. Companies like ATI Technologies, Advance Logic, Chips & Technologies, Headland Technology, Matrox Electronic Systems, and Primus Technology develop many of the accelerator ICs in use today. Many of the ICs provide a 32-bit data bus (even newer designs are providing a 64 bit data bus), and they sustain very high data rates, although a data bottleneck across a 16 bit expansion bus can still seriously degrade the IC's capability. This means you should match the recommended board to the particular system—a state-of-the-art graphics accelerator will not necessarily make your old i286 shine.

Video memory

Video adapters rely on RAM to hold image data, and accelerators are no exception. While the current amount of video RAM typically varies from 512 kbytes to 2 Mbytes (some accelerator boards offer 4 Mbytes), the *amount* of RAM is not as important to a video accelerator as the RAM's *speed*. Faster memory is able to read and write image data faster, so adapter performance is improved. The introduction of specialized video RAM (VRAM)—memory devices with two separate data busses that can be read from and written to simultaneously—is reputed to be superior to conventional dynamic RAM (DRAM) such as the kind used for ordinary PC memory. Recent advances in DRAM speed have narrowed that gap while still remaining very economical. At this point, adapters with fast DRAM are just about as fast as adapters with specialized video RAM for video modes up to $1024 \times 768 \times 256$. For higher modes and color depths found on high-end accelerators, specialized video RAM is still the way to go for optimum performance.

Video BIOS and drivers

Software is often considered as an afterthought to adapter design, yet it plays a surprisingly important role in accelerator performance. Even the finest accelerator board can bog down when run with careless, loosely written code. There are two classes of software that you must be concerned with: video BIOS and drivers. The video BIOS is *firmware* (software that is permanently recorded on a memory device such as a ROM). Video BIOS holds the programming that allows the accelerator to interact with the PC and applications software. VESA BIOS extensions are now being used as part of the video BIOS for many accelerators as well as conventional frame buffer adapters. The addition of VESA BIOS extensions to video BIOS eliminates the need to load another device driver.

However, there are compelling advantages to video drivers. Windows works quite well with drivers (and ignores video BIOS entirely). Unlike BIOS ROMs, which can never change once programmed, a video driver can change very quickly as bugs are corrected and enhancements are made. The driver can be downloaded from a manufacturer's BBS or their forum on CompuServe (or other on-line information service) and installed on your system in a matter of minutes without ever having to disassemble the PC. It is also possible for you to use third-party video drivers. Hardware manufacturers are not always adept at writing efficient software. A third-party driver developed by an organization that specializes in software may actually let your accelerator perform better than the original driver shipped from the manufacturer.

The RAMDAC

Just about every analog video system in service today is modeled after the 15-pin VGA scheme, which uses three separate analog signals to represent the three primary colors. The color for each pixel must be broken down into component red, green, and blue levels, and those levels must be converted into analog equivalents. The conversion from digital values to analog levels is handled by a digital-to-analog converter (DAC). Each conversion also requires a certain amount of time. Faster DACs are needed to support faster horizontal refresh rates. Remember that each video adapter uses a *palette* which is a subset of the colors that can possibly be produced. Even though a monitor may be able to produce an unlimited number of colors, a VGA board can only produce 256 of those colors in any 256 color mode. Older video boards stored the palette entries in registers, but the large-palette video modes now available (64K colors through 16 million colors) require the use of RAM. Boards that incorporate a random access memory digital-to-analog converter (RAMDAC) are preferred because memory integrated with DACs tends to be much faster than accessing discrete RAM elsewhere on the board.

Bus architectures

Finally, graphic data must be transferred between the PC motherboard and the adapter as you saw early in this chapter. Such transfer takes place across the PC's *expansion bus*. If data can be transferred between the PC and adapter at a faster rate, video performance should improve. Consequently, the choice of *bus architecture* has a significant impact on video performance. Video accelerators are available to support three bus architectures: ISA, VL, and PCI.

The venerable *Industry Standard Architecture* (ISA) has remained virtually unchanged since its introduction with the PC/AT in the early 1980s. The ISA continues to be a mature interface standard for most IBM-compatible expansion devices. The sheer volume of ISA systems currently in service guarantees to keep the ISA on desktops for at least another 10 years. However, ISA's 16-bit data bus width, its lack of advanced features such as *interrupt sharing* or *bus mastering,* and its relatively slow 8.33-MHz operating speed form a serious bottleneck to the incredible volume of video data demanded by Windows and many DOS applications. Industry Standard Architecture works, but it is no longer the interface of choice to achieve optimum video performance. When recommending an accelerator product, look to the newer busses for best results.

The *Video Electronics Standards Association* (VESA) has invested a great deal of time and effort to develop a standard bus interface which has been optimized for video operation. In essence, this new video bus is "local" to the system CPU, which allows faster access without the 8.33-MHz limitation imposed by ISA. The actual bus speed is limited by the system clock speed. The VESA local bus (VL bus) has achieved a remarkable level of industry acceptance and success in boosting video performance, especially when used with a high-quality graphics accelerator board. However, the 32-bit VL bus is generally limited to video systems. Other peripherals such as IDE hard drive controllers have been built for the VL bus, but the demands of electrical signals at such high speeds limit the number of VL slots to one or two. Just about every accelerator product on the market today is available in a VL bus implementation. VL accelerators are a safe, inexpensive choice for current systems.

Intel's *Peripheral Component Interconnect* (PCI) bus is one of the newest and most exciting bus architectures to reach the PC. The PCI bus runs at 33 MHz and offers a full 64-bit data bus which can take advantage of new 64-bit CPUs such as Intel's Pentium. While the PCI bus also hopes to overcome the speed and functional limitations of ISA, the PCI architecture is intended to support all types of PC peripherals (not just video boards). Current PCI video boards are now about as expensive as VL adapters and appear to be delivering performance that is roughly equivalent. But as 64-bit CPUs and motherboards become common, it is likely that PCI boards will easily outperform 32-bit VL boards while their prices drop sharply. At the time of this writing, PCI expansion devices are generally limited to video and disk adapters.

Video feature connectors

PC designers have long understood the importance of video performance. The *video feature connector* (VFC) is one example of such forward thinking (Fig. 4-9). Early VGA boards—and many early graphics accelerator designs—employed the VFC. It is the same connector layout that IBM used in their PS/2 display adapter, and many ISA video boards included the connector in an effort to maintain full compatibility with the IBM architecture as well as to support possible future features. The VFC was typically implemented as a 26-pin card-edge connector, though a few video boards provided the connections on an IDC (insulation displacement connector) header. Table 4-3 illustrates the pinout for a classic VFC. Unfortunately, no real use ever developed for the VFC. This is likely due to two reasons. First, most systems have limited expansion space and resources (i.e., interrupts), so there was little market acceptance for a multiboard video system. Second, the explosive growth of gate arrays, application-specific ICs, and powerful video chipsets allowed designers to integrate all needed

■ **4-9** *Diagram of a video feature connector.*

Pin (component side)	Function	Pin (solder side)	Function
Y1	PD0 (DAC pixel data)	Z1	ground
Y2	PD1	Z2	ground
Y3	PD2	Z3	ground
Y4	PD3	Z4	Select Internal Video
Y5	PD4	Z5	Select Internal al Syncs
Y6	PD5	Z6	Select Internal DAC
Y7	PD6	Z7	not used
Y8	PD7	Z8	ground
Y9	DAC Clock	Z9	ground
Y10	DAC Blanking	Z10	ground
Y11	Ext. Horiz Sync	Z11	ground
Y12	Ext. Vert Sync	Z12	not used
Y13	ground	Z13	not used

video functions and memory onto a single board. Still, you should recognize the VFC on sight.

Troubleshooting video adapters

A PC video system consists of four parts: the host PC itself, the video adapter, the monitor, and the software (video BIOS and drivers). To deal with a failure in the video system, you must be able to isolate the problem to one of these four areas. When isolating the problem, your best tool is a working PC. If you do not have a second PC available, perhaps a friend or colleague can make his or her PC available for a brief test.

Isolating the problem area

The first step is to verify the monitor by testing it on an efficiently running PC. Keep in mind that the monitor must be compatible with the video adapter on which it is being tested. If the monitor works on another PC, the fault lies in one of the three remaining areas. If the monitor fails on a good machine, try the good monitor on the questionable machine. If the good monitor then works on your questionable machine, you can be certain that the fault lies with your monitor, and you can refer to the appropriate chapter here for detailed troubleshooting if you wish.

If the monitor checks out, suspect the video adapter. Follow the same process to check the video adapter. Try the suspect video adapter on a known-good PC. If the problem follows the video adapter, you can replace the video adapter. If the suspect video adapter works in a good system, the adapter is probably good. Replace the adapter in the suspect machine, but try another expansion slot and make sure that the monitor cable is attached securely.

If both the monitor and the video adapter work in a good PC, but the video problem persists in the original machine, suspect a problem with the PC motherboard. Either the expansion slot is faulty, or a fault has occurred on the motherboard. Run some PC diagnostics if you have them available. Diagnostics may help pinpoint motherboard problems. You may then choose to troubleshoot the motherboard further or replace the motherboard outright at your discretion. Try the working video adapter in another expansion slot.

When the video system appears to work properly during system initialization but fails with a particular application (or in Windows), strongly suspect a problem with the selected video driver. Since almost all video adapters support VGA at the hardware level, set your application (or change the Windows setup) to run in standard VGA mode. If the display functions properly at that point, you can be confident that the problem is driver related. Check with the manufacturer to see that you have the latest video driver available. Reload the driver from its original disk (or from a new disk) or select a new driver. If the problem persists in VGA mode, the trouble may be in the video adapter.

Component-level repairs

A word about board-level repairs. It takes three things to troubleshoot a circuit: test equipment, schematics/diagrams, and replacement parts. Unfortunately, it is difficult to obtain technical information on most video adapters. Schematics are often guarded jealously by manufacturers. Even if schematics were readily available, many of the highly integrated components used in current video adapters are proprietary, that is, designed and made specifically for a certain manufacturer and not available in the general marketplace. This can make it extremely difficult to locate key replacement parts. Given the relatively low cost for most video boards, it often makes more economic sense simply to replace the board rather than deal with the logistics of troubleshooting it to the component level unless you have a great number of boards to fix and resell. Warranties are another consideration. It is unwise to tinker with a video adapter (or any PC product) while it is still under warranty. Why spend time and resources dealing with a problem that the manufacturer will deal with for free? For the purposes of this book, video adapters will be considered as replaceable parts. Defective adapters should be replaced outright.

Symptoms and solutions

Symptom 1 *The computer is on, but there is no display.* Make sure that the monitor is turned on and plugged into the video adapter properly. Also check that the monitor's brightness and contrast controls are turned up enough (it sounds silly, but it really does happen). Try the monitor on a good PC. If the monitor works properly, suspect the video adapter. Power down the PC and make sure the video adapter is seated properly in its expansion slot. If any of the board contacts are dirty or corroded, clean the contacts by rubbing them with an eraser. You can also use a cotton swab and any electronics-grade contact cleaner. You may want to try the video board in another expansion slot.

Chances are that the video adapter has at least one hardware jumper or DIP switch setting. Contact the manufacturer or refer to the owner's manual for the board and check that any jumpers or DIP switch settings on the board are configured properly. If this is a new installation, check the adapter board settings with the configuration of other expansion boards in the system. When the hardware settings of one board overlap the settings of another, a hardware conflict can result. When you suspect a conflict, change the settings of the video adapter (or another newly installed device) to eliminate the conflict.

There may also be a memory conflict. Some video adapters make unusual demands of upper system memory (the area between 640 kbytes and 1 Mbyte). It is possible that an *Exclude* switch must be added to the EMM386.EXE entry in a CONFIG.SYS file. Check with the adapter's instruction manual to see if there are any memory configuration changes or optimizations that are required for proper operation.

Symptom 2 *There is no display (the computer sounds one long and two short beeps).* You may also encounter a pattern of beep codes (i.e., 3-3-4, 3-4-1, 3-4-2, or 3-4-3). The video adapter failed to initialize during the system's POST. Because the video adapter is not responding, it is impossible to display information, which is why a series of beeps is used. Bear in mind that the actual beep sequence may vary from system to system depending on the type of BIOS being used. In actual practice, there may be several reasons

113

why the video adapter fails. Power down the PC and check that the video adapter is installed properly and securely in an expansion slot. Make sure that the video adapter is not touching any exposed wiring or any other expansion board.

Isolate the video adapter by trying another adapter in the system. If the display works properly with another adapter installed, check the original adapter to see that all settings and jumpers are correct. If the problem persists, the original adapter is probably defective and should be replaced. If a new adapter fails to resolve the problem, there may be a fault elsewhere on the motherboard. Install a POST board in the PC and allow the system to initialize. Each step of the initialization procedure corresponds to a two-digit hexadecimal code shown on the POST card indicators. The last code to be displayed is the point at which the failure occurred. POST cards are handy for checking the motherboard when a low-level fault has occurred. If a motherboard fault is detected, you may troubleshoot the motherboard or replace it outright at your discretion.

Symptom 3 *You see large blank bands at the top and bottom of the display in some screen modes, but not in others.* Multifrequency and multimode monitors sometimes behave this way. This is not necessarily a defect, but it can cause some confusion unless you understand what is going on. When screen resolution changes, the overall number of pixels being displayed also changes. Ideally, a multifrequency monitor should detect the mode change and adjust the vertical screen size to compensate (a feature called *autosizing*). However, not all multifrequency monitors have this feature. When video modes change, you are left to adjust the vertical size manually.

Refer to the user guide for the monitor and see if there is an autosizing feature built in. If there is, check the adapter and see if the feature is enabled. You may need to move a jumper or DIP switch to enable the function. If the monitor will not autosize the display, the display may be defective. If the monitor does not offer autosizing, there is little to be done except to continue adjusting the vertical size manually (or upgrade to an autosizing monitor).

Symptom 4 *The display image rolls.* Vertical synchronization is not keeping the image steady (horizontal sync may also be affected). This problem is typical of a monitor that cannot display a particular screen mode. Mode incompatibility is most common with fixed-frequency monitors, but it can also appear in multifrequency monitors that are being pushed beyond their specifications. The best course of action here is simply to reconfigure your software to use a compatible video mode. If that is an unsatisfactory solution, you will have to upgrade to a monitor that will support the desired video mode.

If the monitor and video board are compatible, there is a synchronization problem. Try the monitor on a good PC. If the monitor also fails on a good PC, try the good monitor on the original PC. If the good monitor works on the suspect PC, the sync circuits in your original monitor have almost certainly failed. If the suspect monitor works on a good PC, the trouble is likely in the original video adapter. Try replacing the video adapter.

Symptom 5 *An error message appears on system startup indicating an invalid system configuration.* The system CMOS backup battery has probably failed. This is typically a symptom that occurs in older systems. If you enter your system setup and examine each entry, you will probably find that all entries have returned to a default setting, including the video system setting. Your best course is to replace the CMOS backup battery and enter each configuration setting again (ideally you have recorded each setting on paper already or saved the CMOS contents to floppy disk using a CMOS backup utility). Once new settings are entered and saved, the system should operate properly. If the CMOS still does not retain system configuration information, the CMOS RAM itself is probably defective. Use a software diagnostic to check the motherboard thoroughly. If a motherboard fault is detected, you can troubleshoot the motherboard or replace it outright at your discretion.

Symptom 6 *Garbage appears on the screen or the system hangs up.* There are a variety of reasons why the display may be distorted. One potential problem is a monitor mismatch. Some adapters must be deliberately set to work with fixed-frequency or multifrequency monitors. Check the video adapter jumpers and DIP switch settings to be sure that the video board will support the type of monitor you are using. It is possible that the video mode in use is not supported by your monitor (the display may also roll as described in Symptom 4). Try reconfiguring your application software to use a compatible video mode. The problem should disappear. If that is an unsatisfactory solution, you will have to upgrade to a monitor that will support the desired video mode.

Some older multifrequency monitors are unable to switch video modes without being turned off and then turned on again. When such monitors experience a change in video mode, they will respond by displaying a distorted image until the monitor is reset. If you have an older monitor, try turning it off, wait several minutes, and then turn it on again.

Conflicts between device drivers and terminate-and-stay-resident (TSR) programs will upset the display and are particularly effective at crashing the computer. The most effective way to check for conflicts is to create a backup copy of your system startup files CONFIG.SYS and AUTOEXEC.BAT. From the root directory (or directory that contains your startup files), type;

```
copy autoexec.bat autoexec.xyz
copy config.sys config.xyz
```

The extensions "xyz" suggest that you use any three letters, but avoid using "bak" because many ASCII text editors create backup files with this extension.

Now that you have backup files, go ahead and use an ASCII text editor (i.e., the text editor included with DOS) to REM-out each driver or TSR command line. Reboot the computer. If the problem disappears, use the ASCII text editor to reenable one REMed-out command at a time. Reboot and check the system after each command line is reenabled. When the problem occurs again, the last command you reenabled is the cause of the conflict. Check that command line carefully. There may be command line switches that you can add to the startup file that will load the driver or TSR without causing a conflict. Otherwise, you would be wise to leave the offending command line REMed-out. If you encounter serious trouble in editing the startup files, you can simply recopy the backup files to the working file names and start again.

Video drivers also play a big part in Windows. If your display problems are occurring in Windows, make sure that you have loaded the proper video driver and that the driver is compatible with the monitor being used. If problems persist in Windows, load the standard generic VGA driver. The generic VGA driver should function properly with virtually every video board and VGA (or SVGA) monitor available. If the problem disappears when using the generic driver setup, the original driver is incorrect, corrupt, or obsolete. Contact the driver manufacturer to obtain a copy of the latest driver version. If the problem persists, the video adapter board may be incompatible with Windows. Try another video adapter.

CRT testing
and alignment

COMPUTER MONITORS ARE NOTORIOUSLY RUGGED DEVICES
(Fig. 5-1). The CRT itself enjoys a reasonably long life span. By their
very nature, CRTs are remarkably tolerant to physical abuse and
can withstand wide variations in power and signal voltages. However, even the best CRT and its associated circuitry suffer eventual
degradation with age and use. Monitor operation can also be upset
when major subassemblies are replaced, such as circuit boards or
deflection assemblies. Maintenance and alignment procedures are
available to evaluate a monitor's performance and allow you to keep
it working within its specifications. This chapter illustrates a comprehensive set of procedures that can be performed to test and adjust the monitor's performance. Keep in mind that this chapter uses
test patterns generated with the companion software diskette available from Dynamic Learning Systems. If you do not already have the
diskette, Chapter 10 shows you how to order and use the software.

Before you begin

Adjusting a monitor is a serious matter and should not be under-
taken without careful consideration. There are a myriad of adjustments found on the main circuit board, any of which can render a
screen image unviewable if adjusted improperly. The delicate
magnets and deflection assemblies around the CRT's neck can
easily be damaged or knocked out of alignment by careless handling. *In short, aligning a monitor can do more harm than
good unless you have the patience to understand the purpose
of each procedure.* You also need to have a calm, methodical approach. The following points may help to keep you out of trouble.

Testing versus alignment

There is a distinct difference between monitor testing and monitor
alignment. *Testing* is a low-level operation. Testing is also unob-

■ **5-1** *An NEC MultiSync 5FGe monitor. (NEC Technologies, Inc.)*

trusive, so you can test a monitor at any point in the repair process. By using the companion software with your PC, you can test any monitor that is compatible with your video adapter. Testing is accomplished by displaying a test pattern on the monitor. After observing the condition of the pattern, you can usually deduce the monitor's fault area very quickly. Chapter 10 discusses the companion software in detail. Once the monitor is working (able to display a steady full-screen image), you can also use test patterns to evaluate the monitor's alignment.

Alignment is a high-level operation, a task that is performed only *after* the monitor has been completely repaired. Since alignment requires you to make adjustments to the monitor circuits and assemblies, the monitor circuits *must* be working properly, which means your repair must be complete. After all, alignment does no good if a fault is preventing the monitor from displaying an image in the first place. Proper alignment is important to ensure that the monitor is displaying images as accurately as possible.

Know the warranty

A warranty is a written promise made by manufacturers that their product will be free of most problems for some period of time. For monitors, typical warranties cover parts and labor for a period of 1 year from the date of purchase. The CRT itself is often covered up to 3 years. Before even touching the monitor, you should check to see if the warranty is still in force. Monitors that are still under warranty should be sent back to the manufacturer for repair. This book does *not* advocate voiding any warranty, and the reasoning is very simple: *Why spend your time and effort to do a job that the manufacturer will do for free?* You (or the monitor's owner) **paid** for that warranty at the time of purchase. Of course, most monitors in need of service are already out of warranty.

There are only three exceptions that might prompt you to ignore a warranty. First, the warranty may already be void if you purchased the monitor "used." Manufacturers typically support the warranty only for the original purchaser (the individual that returns the warranty card). Third-party claims are often refused, but it may still be worth a call to the manufacturer's service manager just to be sure. Second, any warranty is only as good as the manufacturer. A manufacturer that goes out of business is not concerned with supporting your monitor, although reputable manufacturers that close their doors will turn over their service operations to an independent repair house. Again, a bit of detective work may be required to find out if the ultimate service provider will honor the unit's warranty. Finally, you may choose to ignore a warranty for organizations with poor or unclear service performance. Call the service provider and ask about the turnaround time and return procedures. If they "don't know" or you can't get a straight answer, chances are that your monitor is going to sit untouched for quite a while.

Getting from here to there

Monitors require special care in moving and handling. A monitor is typically a heavy device (most of its weight is contributed by the CRT and chassis). Back injury is a serious concern. If you *must* move the monitor between places, remember to lift from the knees and *not* from the back—it's hard to fix a monitor while you're in traction. When carrying the monitor, keep the CRT face toward your chest with your arms wrapped carefully around the enclosure to support the weight. Large monitors (17 to 20 in. and bigger) are particularly unwieldy. You are wise to get another person's help when moving such bulky, expensive devices. If the monitor is to be used in-house, keep it on a roll-around cart so that you will not have to carry it.

When transporting a monitor from place to place in an automobile, the monitor should be sealed in a well-cushioned box. If an appropriate box is not to be had, place the monitor on a car seat that is well-cushioned with soft foam or blankets; even an old pillow or two will due. Set the monitor on its cushioning *face down* since that will lower the monitor's center of gravity and make it as stable as possible. Use twine or thin rope to secure the monitor so it will not shift in transit.

When sending the monitor to a distant location, the monitor should be shipped in its original container and packing materials. If the original shipping material has been discarded, purchase a heavy-gauge cardboard box. The box dimensions should be at least 4 to 6 in. bigger in every dimension than the monitor. Fill the empty space with plenty of foam padding or bubble wrap, which can be obtained from any full-service stationery store. The box should be sealed and reinforced with heavy-gauge box tape. It does not pay to skimp here; a monitor's weight demands a serious level of protection.

High-voltage cautions

It is important to remind you that a computer monitor uses very high voltages for proper operation. Potentially *lethal* shock hazards exist within the monitor assembly both from ordinary ac line voltage as well as from the CRT anode voltage developed by the flyback transformer. You must exercise **extreme** caution whenever the monitor's outer housings are removed. *If you have not yet read about shock hazard dangers and precautions in Chapter 2, please read and understand that material* ***now.***

The mirror trick

Monitor alignment poses a special problem for technicians. You must watch the adjustment that you are moving while also watching the display to see what effect the adjustment is having. Sure, you could watch the display and reach around the back of the monitor, but given the serious shock hazards that exist with exposed monitor circuitry, that is a very unwise tactic (you would place your personal safety at risk). Monitor technicians use an ordinary mirror placed several feet in front of the CRT. That way, you can watch inside the monitor as you make an adjustment and then glance up to see the display reflected in the mirror. If a suitable mirror is not to be found, ask someone to watch the display for you and relate what is happening. Ultimately, the idea is that you should never take your eyes off of your hand(s) while making an adjustment.

Making an adjustment

Monitor adjustments are certainly not difficult to make, but each adjustment should be the result of careful consideration rather than a random, haphazard "shot in the dark." The reason for this concern is simple: It is just as easy to make the display *worse* as it is to adjust it correctly. Changing adjustments indiscriminately can quickly ruin display quality beyond your ability to correct it. The following three guidelines will help you make the most effective adjustments with the greatest probability of improving image quality.

First, *mark* your adjustment (Fig. 5-2). Use a narrow-tip indelible marker to make a reference mark along the body of the adjustment. It does not have to be anything fancy. By making a reference mark, you can quickly return to the exact place you started. A reference mark can really save the day if you get lost or move the wrong adjustment.

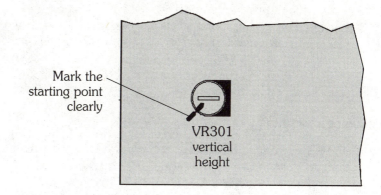

Mark the starting point clearly

VR301 vertical height

■ **5-2** *Mark the starting point before beginning an adjustment.*

Second, concentrate on only *one* adjustment for any one alignment procedure. For example, if you are trying to optimize horizontal linearity, you should *only* be concerned with the horizontal linearity adjustment. If you do not have documentation that describes the location of each control, check the silk-screen labels on the PC board. If you absolutely cannot locate the needed adjustment point, skip the alignment and move on to the next test. When you *do* move an adjustment, move it slowly and in very small increments (perhaps 1/8 to 1/4 turn). Check the display after each step. If the display fails to improve, return the control to its original location (a snap to do if you've made a reference mark) and try it in the opposite direction.

Finally, avoid using metal tools (e.g., screwdrivers) to make your adjustments. Some of the controls in a monitor are based on coils with permeable cores. Inserting steel tools to make an adjustment will throw the setting off—the display may look fine with the tool inserted but degrade when the tool is removed. As a general rule, use plastic tools, such as TV alignment tools, which are available from almost any electronics store.

Tests and procedures

Testing a computer monitor is easy. Simply insert the companion disk in your floppy drive and start it or install and start it from your hard drive. Chapter 10 explains how to obtain, install, and use the software in detail. Once the companion software is started, you can select the test pattern for the specific test you wish to run. Each of the following procedures discusses how to interpret the pattern and provides step-by-step instructions for making adjustments. For the purposes of this discussion, you should refer to the sample main board shown in Fig. 5-3. *Remember that the PC board(s) used in your particular monitor may be quite different, so examine your own PC board very closely before attempting an adjustment.*

High-voltage test and regulation

The high-voltage test is one of the more important tests that you will perform on computer monitors. Excessive high-voltage levels can allow X radiation to escape the CRT. Over long-term exposure, X rays pose a serious biohazard. Your first check should be to use a high-voltage probe similar to the one shown in Chapter 3 (Fig. 3-21). Ground the probe appropriately and insert the metal test tip under the rubber CRT anode cap. You can then read the high-voltage level directly from the probe's meter. *Be certain to refer to any particular operating and safety instructions that accompany your high-voltage probe.* If the high-voltage level is unusually high or unusually low, carefully adjust the level using a high-voltage control (Fig. 5-3 shows VR501 as the high-voltage control), which is usually located near the flyback transformer.

Regulation is the ability of a power supply to provide a constant output as the load's demands change. The high-voltage supply

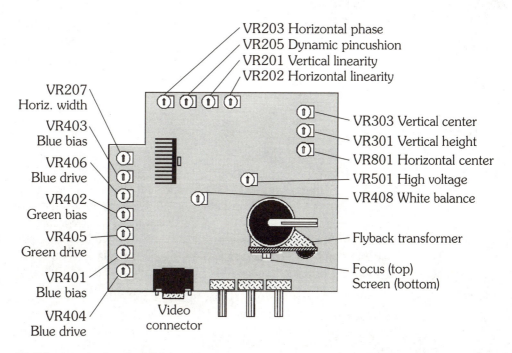

VR203 Horizontal phase
VR205 Dynamic pincushion
VR201 Vertical linearity
VR202 Horizontal linearity

VR207 Horiz. width
VR403 Blue bias
VR406 Blue drive
VR402 Green bias
VR405 Green drive
VR401 Blue bias
VR404 Blue drive

Video connector

VR303 Vertical center
VR301 Vertical height
VR801 Horizontal center
VR501 High voltage
VR408 White balance

Flyback transformer

Focus (top)
Screen (bottom)

■ **5-3** *A typical main PC board layout.*

must also provide regulation within specified limits. As the display image changes, high-voltage levels should remain relatively steady. If not, the display image will flinch as height and width change. With the companion software running, select the *high-voltage test* pattern (Fig. 5-4) from the main menu. This is a narrow white double boarder with a solid white center. Watch the

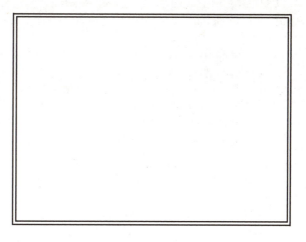

■ **5-4** *The high-voltage regulation test pattern.*

boarder as the center switches on and off at 2 s intervals. A well-regulated high-voltage system set at the correct level will keep the white border reasonably steady; there should be very little variation in image height or width. If the image flinches significantly, the high-voltage system may be damaged or failing.

Screen control

Remember from earlier chapters that the CRT screen grid provides a form of master control over the electron beam(s) which affects the display's overall brightness. A proper screen setting is important so that the brightness and contrast controls work within an appropriate range. Start the companion software (if it is not already running) and select the *blank raster test* from the main menu. Adjust the monitor's *brightness* and *contrast* controls to their maximum levels, and the background raster should be plainly visible.

Locate the *screen voltage* control. In Fig. 5-3, the screen voltage control is located just below the focus control on the flyback transformer assembly. Slowly adjust the screen voltage control until the background raster is just barely visible. Set the monitor's brightness control to its middle (detent) position. The background raster should now be invisible. Press any key to return to the main menu. You may reduce the monitor's contrast control to achieve a clear image.

Focus

When an electron beam is first generated in a CRT, electrons are not directed very well. A focus electrode in the CRT's neck acts to narrow the electron stream. An improperly focused image is difficult to see and can lead to excessive eye strain resulting in headaches, fatigue, and so on. The *focus test* pattern allows you to check the image clarity and optimize the focus if necessary. Keep in mind that focus is a subjective measurement that depends on your perception. You would be wise to confer with another person while making focus adjustments because his or her perception of the display may be different than yours.

Since focus is indirectly related to screen brightness and contrast, you should set the screen controls for an optimum display. An image that is too bright or has poor contrast may adversely affect your perception of focus. Start the companion software if it is not running already. Select the *blank raster test* from the main menu. Adjust screen brightness to its middle (detent) position or until the display's background raster disappears (the screen should be perfectly dark). Press any key to return to the main menu and then select the *white purity test*. The display should be filled with a solid white box. Adjust screen contrast to its maximum position or until a good white image is achieved. Once the display conditions are set properly, press any key to return to the main menu.

Select the *focus test* pattern (Fig. 5-5) from the main menu. You will see a screen filled with the letter *m*. Review the entire screen carefully to determine if the image is out of focus. Again, it is wise to get a second opinion before altering the focus. If the image requires a focus adjustment, gently and slowly alter the focus control. For the sample main board shown in Fig. 5-3, the focus control is located on the flyback transformer assembly. Once you are satisfied with the focus, press any key to return to the main menu.

■ **5-5** *The focus test pattern.*

Dynamic pincushion

A computer-generated image is produced in two dimensions and is essentially flat. Unfortunately, the traditional CRT face is *not* flat (although some new CRT designs use an extremely flat face). When a flat image is projected onto a curved surface, the image becomes distorted. Typically, the edges of the image bow outward making straight lines appear *convex* (barrel distortion). Monitor raster circuitry is designed to compensate for this distortion and allows the image to "appear" flat even though it is being projected onto a slightly curved surface. This is known as the *dynamic pincushion* circuit (or just the *pincushion*). However, if the pincushion circuit overcompensates for curvature, the edges of an image will appear to bow inward making straight lines appear *concave*. Figure 5-6 illustrates the concepts of dynamic pincushion.

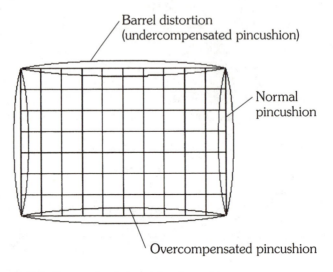

■ **5-6** *Concepts of dynamic pincushion.*

It is a simple matter to check the dynamic pincushion. Start the companion software and select the *convergence test (crosshatch)* pattern from the main menu. A white grid will appear in the display. Inspect the outer boarder of the grid pattern. If the edges of the boarder appear straight and true, the dynamic pincushion is set properly and no further action is needed. If the edges appear to bow outward, the pincushion is undercompensated. If the edges appear to bow inward, the pincushion is overcompensated. In either case, you will need to make a minor adjustment to the dynamic pincushion control. Figure 5-3 lists VR205 as the dynamic pincushion control, but your monitor probably uses different nomenclature. Before making such an adjustment, you may wish to confer with another individual since his or her perception of the display may be different from yours. If you cannot locate the dynamic pincushion control, simply move on to the next test. When you are satisfied with the dynamic pincushion, press any key to return to the main menu.

Horizontal phase

When brightness and contrast are set to their maximum levels, you will see a dim, dark gray rectangle formed around the screen image. This border is part of the *raster,* the overall area of the screen that is hit by the electron beam(s). Ideally, the raster is just slightly larger than the typical image. You are able to control the position of an image within this raster area. This is known as *horizontal phase*. The image should be horizontally centered within the raster area. The term *phase* refers to the amount of delay between the time the horizontal scan (raster) starts and the time when pixel data start. By adjusting this delay, you effectively shift the image left or right in the raster area (which should remain perfectly still).

Start the companion software and select the *phase test* pattern from the main menu. A phase pattern will appear as shown in Fig. 5-7. Set the monitor's brightness and contrast controls to their maximum values; raster should now be visible around the image. Locate the horizontal phase control, which is indicated as VR203 in the sample main board shown in Fig. 5-3. Carefully adjust the horizontal phase control until the image is approximately centered in the raster. This need not be a precise adjustment, but a bit of raster should be visible all around the image. Return the monitor brightness control to its middle (detent) position and reduce the monitor contrast control if necessary to achieve a crisp, clear image. Press any key to return to the main menu.

■ **5-7** *The phase test pattern.*

Horizontal and vertical centering

Now that the image has been centered in the raster, it is time to center the image in the display. Centering ensures that the image is shown evenly so that you can check and adjust linearity later on without the added distortion of an off-center image. Start the companion software and select the *convergence test (crosshatch)* pattern from the main menu. If the image appears well centered, no further action is required, and you can press any key to return to the main menu.

The sample main PC board shown in Fig. 5-3 indicates VR801 as the horizontal centering control. Adjust the centering control so that the image is centered horizontally in the display. Figure 5-3 also shows VR303 as the vertical centering control. Adjust this centering control so that the image is centered vertically in the display. These need not be precise adjustments. Keep in mind that many monitors make their centering controls "user accessible" from the front or rear housings (along with brightness and contrast). When the image is centered to your satisfaction, press any key to return to the main menu.

Horizontal and vertical size (height and width)

Many monitors are capable of displaying more than one video mode. Unless the monitor offers an *autosizing* feature, however, the image will shift in size (especially vertical height) for each different video mode. Now that we have a focused, centered image, it should be set to the proper width and height. Remember that image size depends on CRT size, so you will have to check the specifications for your particular monitor. If you do not have specifications available (or they do not specify image dimensions), you can at least approach a properly proportioned image using the companion software.

Start the companion software and select the *convergence test (crosshatch)* pattern from the main menu. This pattern produces a grid, and each square of the grid should be roughly square. If the image is proportioned correctly, no further action is needed, and you can press any key to return to the main menu. Figure 5-3 uses VR301 to control vertical height. Slowly adjust vertical height until the grid squares are actually about square. The entire grid will be a rectangle which is wider than it is high. If the overall image is too small, you can adjust the horizontal width (VR207 in Fig. 5-3) to make the grid wider and then adjust the vertical height again to keep the grid squares in a square shape. Of course, if the image is too large, you can reverse this procedure to shrink the image. When the image is proportioned correctly, press any key to return to the main menu.

If your monitor is a multimode design and able to display images in several different graphics modes, there may be several independent vertical height adjustments, one for each available mode. You will have to check the test mode you have selected against the vertical control to be sure that the vertical height control you are changing is appropriate for the test mode being used. If you are using a 640 × 480 graphics mode, for example, you should be adjusting the vertical height control for the monitor's 640 × 480 mode. If the monitor offers an autosizing feature that automatically compensates image size for changes in screen mode, there may not be a vertical height adjustment available on the main PC board.

Horizontal and vertical linearity

The concept of *linearity* is often a difficult one to grasp because there are so few real-life examples for us to draw from. Linearity is best related to consistency: Everything should be the same as everything else. For a computer monitor, there must be both horizontal and vertical linearity for an image to appear properly. An image is formed as a series of horizontally scanned lines. Each line should be scanned at the same speed from start to finish. If horizontal scanning speed fluctuates, vertical lines will appear closer together (or father apart) than they actually are. Circles will appear compressed (or elongated) in the horizontal direction. Each horizontal line should be spaced exactly the same vertical distance apart. If the spacing between scanned lines should vary, horizontal lines will appear closer together (or farther apart) than they actually are. Circles will appear compressed (or elongated) in the vertical direction. Any "nonlinearity" will result in some distortion to the image.

Before we begin testing, you should understand that the screen mode will have an effect on the test pattern. When screen modes change, the vertical height of the image will also change. This is especially prevalent in multifrequency and multimode monitor designs which can display images in more than one graphics mode. Before testing for linearity, the height and width of the image should be set properly for the selected screen mode as described in the previous procedures. Otherwise, the image may appear compressed or elongated and result in a false diagnosis.

Start the companion software and select the *linearity test (lines & circles)* from the main menu. A grid will appear with an array of five circles as shown in Fig. 5-8. Observe the test pattern carefully. The spacing between each horizontal line should be equal, as should the spacing between each vertical line. Assuming the vertical height and horizontal width are set properly, each of the five

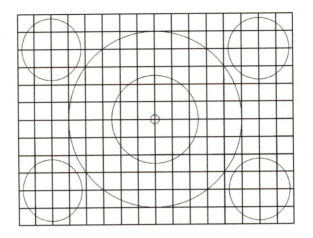

■ **5-8** *The linearity test pattern.*

circles should appear round and even. Each grid square should appear square. If the image appears correct, no further action is needed, so press any key to return to the main menu.

If the vertical lines are not spaced evenly apart, there may be a horizontal linearity problem. Find the horizontal linearity adjustment (VR202 in Fig. 5-3). Be sure to mark the starting point and then slowly adjust the control until the horizontal linearity improves. If there is no improvement in one direction (or linearity worsens), return to the starting point and try the adjustment in the opposite direction. If there is still no improvement (or linearity worsens again), return the control to its starting position and take no further action. There may be a fault in the horizontal drive circuit. Refer to Chapters 7 and 8 on monitor troubleshooting for detailed service procedures.

If the horizontal lines are not spaced evenly apart, there may be a vertical linearity problem. Find the vertical linearity adjustment (VR201 in Fig. 5-3). Mark the starting point and slowly adjust the control until the vertical linearity improves. If there is no improvement in one direction (or linearity worsens), return to the starting point and try the adjustment in the opposite direction. If there is still no improvement (or linearity worsens again), return the control to its starting position and take no further action. There may be a fault in the vertical drive circuit. Refer to the chapters on monitor troubleshooting for detailed service procedures. For now, press any key to return to the main menu.

Static convergence

Convergence is a concept that relates expressly to color CRTs. A color CRT produces three electron beams, one for each of the primary colors (red, green, and blue). These electron beams strike color phosphors on the CRT face. By adjusting the intensity of each electron beam, any color can be produced, including white. The three electron beams must converge at the *shadow mask* which is mounted just behind the phosphor layer. The shadow mask maintains color purity by allowing the beams to impinge only where needed (any stray or misdirected electrons are physically blocked). Without the shadow mask, stray electrons could excite adjacent color phosphors and result in strange or unsteady colors. If the beams are not aligned properly, a beam may pass through an adjacent mask aperture and excite an undesired color dot. Proper convergence is important for a quality color display.

It is a simple matter to check convergence. Be certain to allow at least 15 min for the monitor to warm up. Start the companion software and select the *convergence test (dots)* from the main menu. An array of white dots should appear on the display. Observe the dots carefully. If you can see any "shadows" of red, green, or blue around the dots, convergence alignment may be necessary. If the dot pattern looks good, press any key, return to the main menu and then select the *convergence test (crosshatch)* pattern. A white grid should appear. Once again, observe the display carefully to locate any primary color "shadows" that may appear around the white lines. If the crosshatch pattern looks good, no further action is necessary, so press any key to return to the main menu.

When you determine that a convergence alignment is necessary, be sure to select the *convergence test (crosshatch)* from the main menu. Locate the *convergence rings* located on the CRT's neck just behind the deflection yokes as shown in Fig. 5-9. Using a fine-tip black marker, mark the starting position of each convergence ring relative to the glass CRT neck. This is a vital step that will allow you to quickly return the rings to their original positions if you run into trouble. Convergence alignment is delicate, so it is easy to make the display much worse if you are not *very* careful. Also, this alignment must be performed with monitor power applied, so be **extremely** careful to protect yourself from shock hazards. The alignment process is not difficult, but it requires a bit of practice and patience to become proficient. As the following procedure shows, you will align the red and blue electron beams to

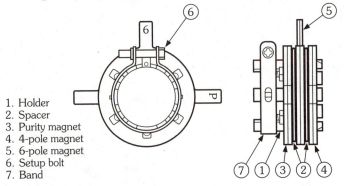

Convergence purity magnet

1. Holder
2. Spacer
3. Purity magnet
4. 4-pole magnet
5. 6-pole magnet
6. Setup bolt
7. Band

■ **5-9** *Identifying the CRT convergence rings. (Courtesy of Tandy Corporation)*

make magenta; then you will align the green electron beam over the magenta pattern to make white.

While the crosshatch pattern is displayed, press the letter *M,* which will switch the crosshatch pattern to a magenta color. Magenta is a combination of blue and red (by choosing magenta, the green electron beam is shut down, so there is less clutter in the display). Loosen the metal band holding the rings in place. *Do not remove the band.* You may also have to loosen a locking ring before moving the convergence rings. Move the *magenta convergence rings* (often referred to as the four-pole magnet) together or individually until the blue and red shadows overlap to form a uniform magenta crosshatch pattern. Be sure to move these rings only in small, careful steps. By moving the rings *together,* you adjust red and blue overlap in the vertical lines. Figure 5-10 shows that by moving the rings *individually* you adjust red and blue overlap in the horizontal lines. If you get into trouble, use the starting position marks to return the rings to their original locations and start again.

Once the magenta pattern is aligned, press the letter *W,* which will switch the crosshatch pattern to its original white color (thus activating the green electron beam). Adjust the *white convergence rings* (often called the O-field, green, or six-pole magnet) until any green shadows overlap the crosshatch to form a uniform white grid as desired. As with the magenta rings, moving the white convergence rings *together* will adjust green overlap in the vertical lines. Moving the white convergence rings *individually* will adjust green overlap in the horizontal lines. When the image appears white, carefully secure any locking ring and setup band. Recheck

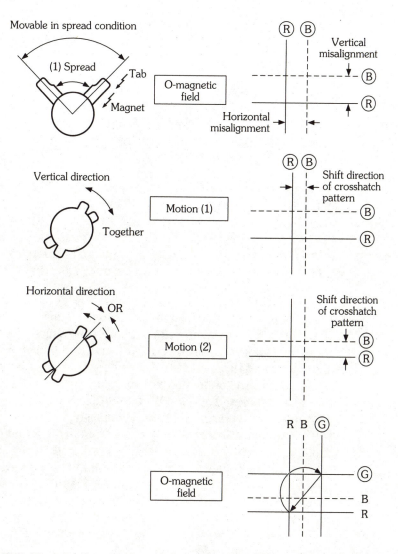

■ 5-10 *Adjusting the four-pole and six-pole convergence magnets. (Courtesy of Tandy Corporation)*

the convergence as you tighten the assembly to be sure nothing has shifted. Do not overtighten the setup band—you stand a good chance of damaging the CRT.

Dynamic convergence

You will probably hear convergence referred to as "static" and "dynamic." These terms refer to the convergence in different areas of the display. *Static* refers to the convergence in the center area of the display. *Dynamic* refers to the convergence around the perimeter of the display. While static convergence provides a good overall alignment with a minimum of fuss, dynamic convergence is a more difficult alignment that requires inserting rubber wedges between the edge of the deflection yoke(s) and the CRT funnel. These are touchy procedures even for a practiced hand. If there is visible misconvergence around the display perimeter even after a careful static convergence alignment, you will have to consider a dynamic convergence procedure.

As you might expect, dynamic convergence is perhaps the most unforgiving alignment procedure. Once you remove the wedges or alter their positions, it is extremely difficult to restore them (even with alignment marks). Unfortunately, there are no formal procedures for positioning the wedges, so you are often left to your own trial-and-error calls. In the end, you should avoid dynamic convergence adjustments if possible.

Dynamic convergence alignment requires several important steps. First, start the white convergence (crosshatch) pattern and allow the monitor to warm up for at least 15 minutes. Then mark and remove all three wedges (make sure they are free to move). When you tilt the deflection yoke up and down, you will see distortion as illustrated in the A portion of Fig. 5-11. As you see, misconvergence increases near the screen edge. Use the first two wedges to align the up/down positioning of the deflection yoke, and the distortion shown in Fig. 5-11 should disappear. Next, when you tilt the deflection yoke right and left, you will see distortion as illustrated in the B portion of Fig. 5-11. Use the third wedge to eliminate that distortion. You may have to tweak the three wedges to optimize the dynamic convergence. Finally, you should secure the wedges in place with a dab of high-temperature–high-voltage epoxy.

The problem with dynamic convergence is that you often make the display worse when removing the wedges (even with installation marks) because you typically have to break the hold of any epoxy and wrench the wedges free. As a consequence, it is extremely difficult to just "tweak" an existing dynamic convergence calibration; in virtually all cases, it is an all-or-nothing proposition. With this in mind, you should evaluate the need for dynamic convergence very carefully before proceeding, and if you do choose to proceed, make sure to leave yourself plenty of time.

Dynamic convergence A

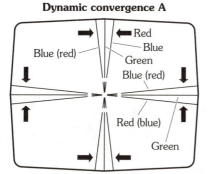

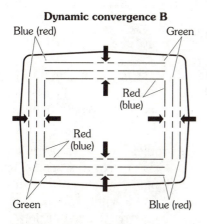

■ **5-11**

Dynamic convergence illustrations.
(Courtesy of Tandy Corporation)

Color purity

Another concern is color *purity,* that is, a solid color should have the same hue across the entire display. If discoloration develops in the display, purity may need to be restored by degaussing (demagnetizing) the monitor. Typically, the discoloration follows a semicircular pattern around one side or corner with several bands of different color distortions (Fig. 5-12), and the color banding appears almost like a rainbow. Sometimes the discoloration involves the entire screen, but that is rare.

Such discoloration may be caused by an externally induced magnetic field that has permanently magnetized some material in the monitor. The three CRT electron beams are guided to their appropriate phosphor dots by a magnetic deflection system. The beams converge and pass through a shadow mask near the phosphor surface assuring that the red beam hits the red phosphor, the blue beam hits the blue phosphor, and the green beam hits the green phosphor. If some component within the CRT (frequently it is the shadow mask itself) has become sufficiently magnetized, then the

135

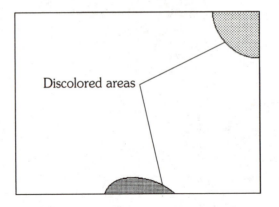

Discolored areas

■ **5-12** *Color purity problems due to shadow mask magnetization.*

beams receive an undesired deflection and will not land on the appropriate phosphor (or will land partly on one color and partly on another). The result is an impure color that arcs around the magnetized area.

Location is an important clue to this problem. If the discoloration moves or disappears when the monitor is moved, it is not being caused by a permanent magnetization, but by some magnetic interference in or near the monitor. Placing a highly magnetic or electromagnetic source (e.g., a strong industrial magnet or power supply) on or near the monitor can cause such discoloration. If the discoloration does not move when the monitor is moved, it may be caused by permanent magnetization in the shadow mask. In that case, degaussing is necessary.

Checking color purity is a straightforward procedure. Start the companion software and select the *white purity test* from the main menu. A white box will fill the entire screen. If there are any areas of discoloration, degaussing is probably necessary. Degaussing removes permanent magnetization by introducing an alternating magnetic field that is stronger than the offending permanent magnetization. This field will energize the magnetic domains of the material and induce an alternating magnetic field. Then, if the amplitude of the alternating magnetic field is gradually reduced to zero, the magnetic domains in the material will be left disorganized and scrambled, and this effectively demagnetizes the monitor.

The easiest way to degauss a monitor is often to let the monitor do it. All modern color monitors have built-in degaussing coils and circuits. There will be a thick black coil of wire wrapped in tape or other insulation surrounding the CRT faceplate. Usually, it is coiled around the CRT behind its mounting ears. That is the internal *degaussing coil*. The coil is connected to the ac supply through a thermistor current limiting circuit. The thermistor has a low resistance when cold and a higher resistance when warm (typically a 10:1 ratio). It is in series with the degaussing coil so that when started cold a large current will flow through the coil and then will decrease to a low value. The internal degaussing coil thus automatically degausses the monitor every time it is turned on. This degaussing takes place while the monitor screen is blank (the video system has not yet initialized) so that the resulting discoloration during autodegaussing is not visible. Unfortunately, design limitations reduce the magnetic field strength available from internal degaussing coils. That limits the amount of permanent magnetization that can be neutralized by internal degaussing. If a monitor has been strongly magnetized, internal degaussing may not be enough, and discoloration eventually results.

Manual degaussing requires a hand-held degaussing coil. You may have to search a bit to find one, but they are available. The basic principle involved in operating a manual degaussing coil is the same as the autodegaussing assemblies already in place on color monitors: Introduce a strong alternating magnetic field and then slowly reduce its amplitude to zero. Start the companion software and select the *white purity test* from the main menu. A white rectangle should fill the entire image. Discoloration should be visible. Hold the degaussing coil near the monitor, flip the degaussing coil switch on, and slowly move the coil away from the monitor as smoothly as you can. The image will discolor drastically when the degaussing coil is activated. When the coil is at arm's length from the monitor, flip the degaussing switch off. You may need to repeat this procedure several times. When the monitor is degaussed properly, the white image should be consistent at all points on the display.

Color drive

Once color convergence and purity are set correctly, you should turn your attention to the color drive levels (also known as *white balance*). Start the companion software and select the *white purity test* from the main menu. A white box will fill the entire screen. Set screen contrast to its highest level and reduce brightness to its middle (detent) position. The background raster should disappear. Ideally, all three color signal levels should be equal, and the resulting image should be a pure white, rather like a blank piece of white photocopier paper. However, judging the quality of a display color is largely a subjective evaluation. You will need an oscilloscope to measure the actual voltage level of each color signal in order to set them equally. If you do not have an oscilloscope (or do not have access to one), do not attempt to adjust the color drive settings "by eye."

Use your oscilloscope to measure the signal levels being generated by the red, green, and blue video drivers. These are the three color signals that are actually driving the CRT. With the full white pattern being displayed, all three color signals should be equal (probably around 30 V, although your own monitor may use slightly different signal levels). Even if you are not quite sure what the level should be, all three signals must be set to the *same* level to ensure a white image. If you do not know what the level should be, find the highest of the three color signals and use that as a reference. Adjust the gain levels of the other two colors until both levels match the reference. Reduce contrast and inspect the image again. It should remain white (and all three signals should be equal). Disconnect your oscilloscope and press any key to return to the main menu.

Cleaning/vacuuming

Once the monitor is checked and aligned, your final step before returning the unit to service should be to inspect the housings and PC boards for accumulations of dust and debris. Look for dust accumulating in the housing vents. A monitor is typically cooled by convection (hot air rises). If these vents become clogged, heat will build up inside the monitor and lead to operational problems and perhaps even cause a premature breakdown. Dust is also conductive. If enough dust builds up within the monitor, the dust may short circuit two or more components and cause operational problems. Vacuum away any dust or debris that may have accumulated in the outer housing. When you see dust buildup around the monitor PC boards and CRT, turn off and unplug the monitor; then vacuum away any buildups. Carefully reassemble the monitor's housing(s) and return it to service.

Power supply troubleshooting

MONITORS REQUIRE SUBSTANTIAL AMOUNTS OF POWER IN order to function properly (Fig. 6-1). This is largely due to the use of a vacuum tube (the CRT itself). Unfortunately, conventional ac is not *directly* compatible with your computer monitor, so the line power from an ac wall outlet must be converted into values of voltage and current that *are* suitable. Such manipulation is the task of the *power supply.* A typical monitor power supply will provide five outputs: 6.3 V to power the CRT cathode, 12 V and 20 V to power the monitor's raster and video circuitry, 87 V to provide the CRT's screen voltage, and 135 V to run the high-voltage system. This chapter presents the concepts and troubleshooting procedures for linear and switching power supplies which are used quite commonly in monitor designs. The chapter also covers the operation and troubleshooting of high-voltage circuits. Finally, there is a section that deals with backlight power supplies for anyone working with flat-panel displays.

Before beginning, you *must* be aware of the hazards involved with ac-powered and high-voltage supply circuits. There will be several points in your monitor where potentially dangerous (even lethal) voltage levels exist. If you have not read about high-voltage hazards yet, refer to the ***Warnings, Cautions, and Human Factors*** section of Chapter 2 before continuing. Be certain to take **all** precautions to protect yourself from injury.

Linear power supplies

The term *linear* means "line" or "straight." As you can see from the block diagram of Fig. 6-2, a linear supply essentially operates in a straight line from ac input to dc output(s). The exact component parts and supply specifications will vary greatly from supply to supply, but *all* linear supplies contain the same basic subsections: a transformer, a rectifier, a filter, and a regulator.

Transformers

A *transformer* constitutes the vast majority of a power supply's weight and overall size. Transformers use the principles of *magnetic coupling* to alter the ac input voltage and current levels. The transformer's output (secondary) is not directly connected to its input. Instead, the output signal is generated by magnetism induced by fluctuations of the ac at the transformer's primary (input). By altering the proportion of primary versus secondary windings, it is possible to convert an ac input signal into a higher (step-up) or lower (step-down) level. Current is also stepped, but

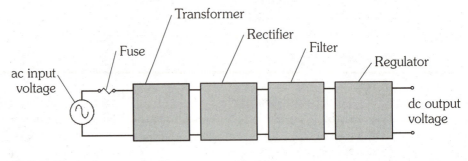

■ **6-2** *Block diagram of a linear power supply.*

in a proportion *opposite* to that of voltage. For example, it is often desirable to transform 120 V ac line voltage into one or more lower levels of ac voltage (i.e., 27 or 18 V ac). Such transformation requires a *step-down* transformer. By stepping the line voltage down, current is stepped up by the same factor. Suppose a 10:1 step-down transformer reduces 120 to 12 V ac. An input current of 100 mA (0.10 A) would be multiplied to 1000 mA (1.00 A).

Rectifiers

The secondary (output) voltage from a transformer is still in ac form. Alternating current voltage and current *must* be converted into dc before powering electronic circuits. The process of converting ac into dc is known as *rectification.* To achieve rectification, only *one* polarity of the ac signal is allowed to reach the rectifier's output. Even though the rectifier's output can vary greatly, the output's polarity will remain either positive or negative. This fluctuating dc is called *pulsating dc.* Diodes are ideal for use as rectifiers because they allow current to flow in only one direction. You may encounter any of three classical rectifier circuits: half-wave, full-wave, and bridge.

A *half-wave rectifier* is shown in Fig. 6-3. It is the simplest and most straightforward type of rectifier since only one diode is required. As secondary voltage from the transformer exceeds the diode's turn-on voltage (about 0.6 V), the diode begins to conduct current. This condition generates an output which mimics the positive half of the ac signal. If the diode were reversed, the output polarity would also be reversed. The disadvantage with half-wave rectifiers is that they are very inefficient. Only half of the ac wave is handled, and the other half is basically ignored and wasted. The resulting gap between pulses causes a lower average output voltage

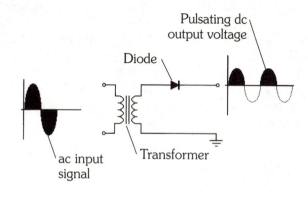

Pulsating dc output voltage

Diode

ac input signal

Transformer

■ **6-3** *Diagram of a half-wave rectifier.*

and a higher amount of ac noise contained in the final dc output. Half-wave rectifiers are rarely used in modern linear power supplies, although switching supplies sometimes use them for final-stage rectification.

Full-wave rectifiers such as the one shown in Fig. 6-4 offer substantial advantages over the half-wave design. By using two diodes in the configuration shown, *both* polarities of the ac secondary voltage can be rectified into pulsating dc. Since a diode is at each terminal of the secondary signal, polarities at each diode are opposite. When the ac signal is negative, the upper diode conducts and the lower diode is cutoff. When the ac signal is negative, the upper diode is cutoff and the lower diode conducts. This means that there is always *one* diode conducting, so there are no gaps in the pulsating dc signal. The only disadvantage of full-wave rectifiers is that a center-tapped transformer is needed. Tapped transformers are often heavier and bulkier than nontapped transformers.

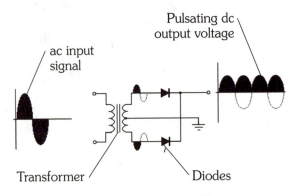

■ 6-4 *Diagram of a full-wave rectifier.*

Diode *bridge rectifiers* use four diodes in a bridge configuration to provide full-wave rectification without the hassle of a center-tapped transformer. Figure 6-5 illustrates a typical bridge rectifier. Two diodes provide forward current paths for rectification, while the other two diodes supply isolation to ground. When ac voltage is positive, diode D1 conducts and D4 provides isolation. When ac voltage is negative, diode D2 conducts while D3 provides isolation. Bridge rectifiers are by far the most popular type of rectifier circuit for linear power supplies.

Filters

By strict technical definition, pulsating dc *is* dc because its polarity remains constant (even if its magnitude changes periodically). Unfortunately, pulsating dc is unsuitable for any type of electronic

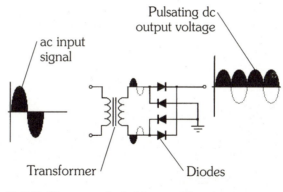

■ 6-5 *Diagram of a bridge rectifier.*

power source. Voltage magnitude must be constant over time in order to operate electronic devices properly. Pulsating dc is converted into smoothed dc through the use of a *filter* as illustrated in Fig. 6-6.

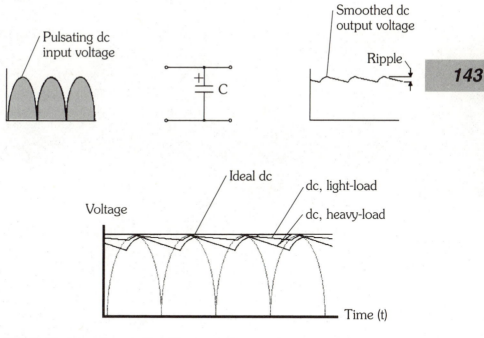

■ 6-6 *Diagram of capacitive filtering action.*

Electrolytic capacitors typically serve as the filter elements since they act as voltage storage devices. When pulsating dc is applied to a capacitive filter, the capacitor charges and voltage across the capacitor increases. Ultimately, the capacitor's charge reaches the peak value of pulsating dc. When a dc pulse falls off back toward zero, the capacitor continues to supply current to the load. This action tends to hold up the output voltage over time—dc is filtered.

However, filtering is not a perfect process. As current is drained away from a filter by the load, voltage across the filter decreases. Filter voltage continues to drop until a new pulse of dc from the rectifier recharges the filter for another cycle. This action of repetitive charging and discharging results in regular fluctuations in the filter output. These fluctuations are called *ripple,* which is an undesirable (but unavoidable) noise component of a smoothed dc output.

The diagram of Fig. 6-6 also shows a sample plot of voltage versus time for a typical filter circuit. The ideal dc output would be a steady line indicating a constant dc output at all points in time. In reality, there will always be some amount of filter ripple. Just how much ripple is present depends on the load being supplied. For a light load (a circuit drawing little current), discharge is less between pulses, so the magnitude of ripple is lower. A large load (a circuit drawing substantial current) requires greater amounts of current, so discharge is deeper between pulses. This results in greater magnitudes of ripple. The relationship of dc pulses is shown for reference.

Beware of shock hazards from power supply filters. Large electrolytic capacitors tend to accumulate substantial amounts of charge and hold that charge for a long time, even after the power supply is turned off. If your fingers or hands touch the leads of a charged capacitor, it may discharge through you. While a capacitor shock is rarely dangerous, it can be very uncomfortable and perhaps result in a mild burn.

Before working on a power supply, be certain that the capacitor is discharged completely by placing a large-value resistor across the capacitor as shown in Fig. 6-7. This *bleeder* resistor slowly drains off any charge remaining on the filter once the power supply is turned off. *Never* attempt to discharge a capacitor by shorting its leads with a screwdriver blade or wire. The sudden release of energy can actually weld a wire or screwdriver directly to the capacitor's leads as well as damage the capacitor internally. Keep in mind that some power supply designs use a *load resistor* across the supply output that will automatically discharge the power supply once power is turned off.

Regulators

A transformer, rectifier, and filter are essential in every power supply. These parts combined will successfully convert ac into dc that

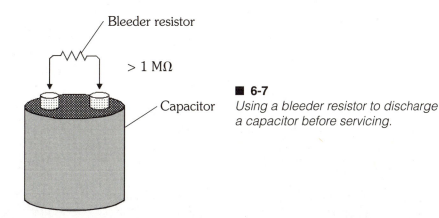

■ 6-7
Using a bleeder resistor to discharge a capacitor before servicing.

Bleeder resistor

> 1 MΩ

Capacitor

is capable of driving many electrical and electronic loads. However, such *unregulated* power supplies have two major disadvantages: Ripple is always present on the supply's output, and the output level varies with the load. Even the most forgiving integrated circuits (ICs) can perform erratically if they are operated with unregulated dc. Ideal dc should be ripplefree and constant regardless of load. A *regulator* is needed to fix dc from a filter's output.

Linear regulation is just as the name implies—current flows from the regulator's input to output as shown in Fig. 6-8. When voltage is applied to the regulator's input, internal circuitry within the regulator manipulates input voltage to provide a steady, consistent output voltage. The output will remain steady under a wide range of input conditions as long as the input voltage is *above* the desired output voltage (often by 3 V or more). If input voltage falls to or below the desired output voltage, the regulator falls out of regulation. In such a case, the regulator's output signal tends to follow the input signal, including ripple.

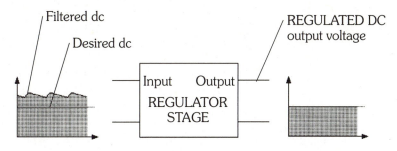

Filtered dc

Desired dc

REGULATED DC output voltage

Input   Output
REGULATOR
STAGE

■ 6-8 *Typical regulator action.*

A linear regulator works by "throwing away" the extra energy provided by a filter; whatever is left over is the desired output. Energy is discarded in the form of heat. This explains why so many regulators are attached to large metal heat sinks. Linear regulation is a simple and reliable method of operation, but it is also very inefficient. Typical linear power supplies are only up to 50 percent efficient. This means that for every 10 W of power generated in the supply, only 5 W are provided to the load. Most of this waste occurs in the regulation process itself.

Zener diodes make excellent voltage regulators because they limit the amount of voltage that occurs across them as illustrated in Fig. 6-9. Input voltage from the filter is applied across the zener diode through a current-limiting resistor. The zener clamps voltage to its zener level. In turn, the zener potential turns on the power transistor, which allows current to flow from collector to emitter. Output voltage will equal the base voltage (established by the zener diode) minus the diode drop across the transistor's base-emitter junction (usually about +0.6 V dc). You can change the regulator's output voltage by changing the zener diode.

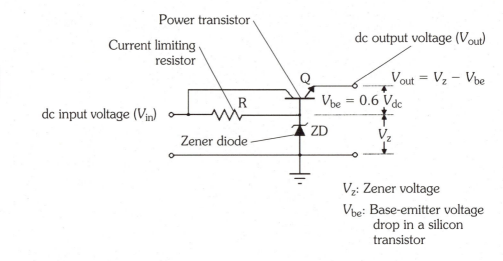

■ **6-9** *Schematic diagram of a simple transistor series regulator.*

Regulator circuits can also be fabricated as integrated circuits as shown in Fig. 6-10. Additional performance features such as automatic current limiting and over-temperature shutdown circuitry can be included to improve the regulator's reliability. Input voltage must exceed the desired output by several volts, but IC regulators are simple to use. One additional consideration for IC regulators is the use of small-value capacitors at the IC's input and output. Small capacitors (0.01 to 0.1 μF) filter high-frequency noise or signals that could interfere with the regulator's operation.

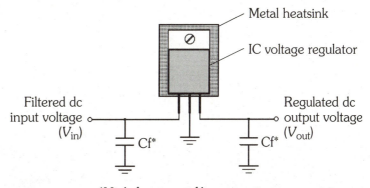

*High-frequency filter capacitors

■ **6-10** *Partial schematic of an IC voltage regulator.*

Troubleshooting linear power supplies

Under most circumstances, linear power supplies are reasonably simple and straightforward to troubleshoot. You can usually make use of your multimeter or oscilloscope to trace voltage through the supply. The point at which your expected voltage disappears is probably the point of failure.

Warning: Keep in mind that many of the following procedures must be performed on **powered** circuitry. Take every precaution to protect yourself and your equipment **before** beginning your repair. The use of an isolation transformer to provide ac is **highly** recommended. If you have not read about voltage cautions in the ***Warnings, Cautions, and Human Factors*** section of Chapter 2, familiarize yourself with that section **now** before continuing.

Start your repair by removing all power from the supply. Disassemble the power supply enough to expose the power supply circuit. Some monitors use a separate PC board for the power supply, whereas others incorporate the power supply on the main PC board. Be certain to insulate any loose assemblies or circuits to prevent accidental short circuits or physical damage. For the following procedures, refer to the example linear power supply shown in Fig. 6-11.

Symptom 1 *The monitor is dead. There is no raster and no picture.* When the monitor is completely inoperative, you should immediately suspect a fault in one or more of the power supply outputs. Begin by checking the ac line voltage into the power supply. Use your multimeter to measure ac voltage at your power supply's plug. If you are using an isolation transformer, you should check the transformer's ac output to your supply. You should nor-

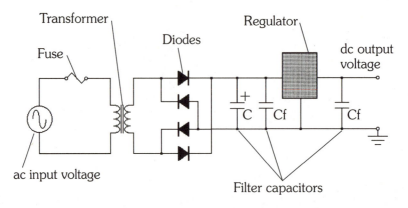

Schematic diagram of a complete linear power supply.

mally read 105 to 130 V ac (210 to 240 V ac in Europe). If your monitor is designed for multinational operation, make sure that the ac voltage selector switch (next to the ac connection) is set appropriately. More or less line voltage can cause the supply to malfunction. Use extreme caution whenever measuring ac line voltage levels.

Check the monitor's power switch to be sure that it is in the on position. This may sound silly, but it really *is* a common oversight. Unplug the power supply and inspect the main line fuse. The fuse is usually accessible from the monitor's outer housing. If there is no external fuse, you must disassemble the monitor to check the fuse(s) that would be located on the power supply circuit. You can test a fuse by removing it from the monitor and measuring its continuity with a multimeter. A working fuse should register as a complete short circuit (0 Ω). If you read infinite resistance, the fuse is defective and should be replaced. Replace defective fuses *only* with fuses of equal size and current rating.

With the monitor unplugged, check all connectors and wiring to be sure that everything is intact. A loose connector or wire can easily disable the supply. If your supply has failed after being dropped or abused, consider the possibility of printed circuit board damage. Faulty soldering at the factory (or on your test bench) can also cause PC board problems. The power PC board may be cracked, one or more component leads may have pulled away from their solder pads, or a trace (or traces) may have broken from impact stress. Inspect the PC board carefully for signs of damage.

Apply power to the monitor and use your multimeter to measure the dc output(s) from the supply. Each output is generally well marked, so you should have no trouble determining what the output should be. When an output is low or zero, disconnect the power supply from its load (if possible) and measure its output(s) again. If you find the same measurements with the load disconnected, the trouble is probably in your supply. If the output returns to its rated value, the supply is probably being shorted by the load circuit(s). Check the monitor circuits for damage or short circuits that may be pulling down the supply output(s).

If there are no obvious failures up to this point, you can try replacing the power supply outright. If the supply is fabricated on its own PC board, you need only replace the power supply assembly. If the supply is incorporated on the main PC board, you have to replace the entire main PC board. You may also choose to troubleshoot the supply circuit to the component level. For the purposes of this discussion, you should follow the procedure with Fig. 6-11. In linear supply troubleshooting, it is often best to start your tracing at the supply output and work backward through the supply toward the ac input.

Measure dc voltage at the regulator's output and input. Regulator output should equal the final supply output measured earlier. If the regulator's input voltage is several volts higher than the expected output and the actual output is low or nonexistent, your regulator is probably defective. Try replacing the regulator. If the regulator's input voltage is low or absent, examine the filter network. Assuming the PC board is intact, the filter voltage should roughly equal the regulator's input voltage. Remove all power from the supply, discharge the filter capacitor(s) with a high-value resistor (100 kΩ or larger), and inspect each capacitor as outlined in Chapter 3. Replace any faulty filter capacitors. If the filter circuit checks properly, inspect the junction of each rectifier diode. An open rectifier diode can disable the supply. Replace any faulty rectifier diodes.

Should your rectifier diodes check properly, reapply power to the supply and measure ac voltage across the transformer's secondary (output) winding. The expected output is typically marked on the windings. Also check the ac voltage across the transformer's primary coil. You should find about 120 V ac across the primary (or about 220 V ac in Europe). If there is no voltage across the primary, there will be no signal across the secondary. This suggests a faulty fuse or circuit breaker, or some other circuit interruption in the primary transformer circuit. When primary voltage is normal and secondary voltage is low or absent, suspect a failure in the transformer. Finally, if ac is available to the supply, but you cannot find the point where ac or dc disappears, you can simply replace the power supply outright.

Symptom 2 *The monitor operates only intermittently. You might see the raster and power indicator LED blink on and off.* An intermittent problem has likely developed in the power supply. Begin by checking the ac input voltage into the power supply. Exercise extreme caution whenever measuring ac line voltage. Use your multimeter to measure ac voltage at your power supply's plug. If you are using an isolation transformer as recommended, you should check the transformer's ac output to your supply. You should normally read 105 to 130 V ac (210 to 240 V ac in Europe). More or less line voltage can cause the supply to malfunction.

Use a multimeter and measure each power supply output before and after an intermittent failure. If any of the supply outputs quit during an intermittent fault, the problem is likely in the power supply. If each of the supply outputs remains steady during an intermittent fault, the supply is probably working properly and the problem is likely elsewhere (e.g., the raster or high-voltage circuits).

When the power supply is suspect, consider the possibility of a PC board failure, especially if the monitor has only recently been dropped or severely abused. Faulty soldering at the factory (or on your test bench) can also cause PC board problems. The PC board may be cracked, one or more component leads may have pulled away from their solder pads, or a trace (or traces) may have broken from the stress of impact. Inspect the power supply PC board assembly carefully and repair any damage that you may find. If the board damage is very extensive or you are unable to find any damage, you should probably replace the power supply outright since intermittent problems are difficult to track down unless they are obvious.

In addition to physical intermittents, you should also check the supply for *thermal* intermittent problems. Thermal problems typically occur in semiconductor devices such as transistors or ICs, so your supply's regulator(s) are likely candidates. A thermal failure is usually indicated when the supply works once turned on, but cuts out after some period of operation. The supply may then remain disabled until it is turned off, or it may cut in and out while running. Often, a thermally defective regulator may operate when cool (room temperature), but as it runs and dissipates power, its internal temperature climbs. When temperature climbs enough, the device may stop working. If you remove power from the supply and let it cool again, the supply may resume operation.

If you detect any hot components in your intermittent supply, you may suspect a thermal intermittent problem. Spray the suspect

150

part(s) with electronics-grade refrigerant available from almost any electronics store. Leave power applied to the part and spray in short, very controlled bursts. Many short bursts are cleaner and more effective than one long burst. If the supply stabilizes or stops cutting out, you have probably identified the faulty part. Replace any thermally intermittent components or replace the entire power supply at your discretion.

Symptom 3 *The main ac fuse fails, and the new replacement fuse also fails.* Remember that a fuse is a protection device. When a fuse opens, it means that excessive current has been drawn into the power supply. Often, this can happen for perfectly innocent reasons such as a power surge occurs on the ac line or the fuse gives out from old age. This will shut down the power supply (and monitor) completely, and if you have read Symptom 1 carefully, you will probably find the defective fuse in a matter of minutes. If the failure was just spontaneous, you can replace the fuse, and the monitor will continue to operate normally (though you should still check it thoroughly just to be safe). However, if the replacement fuse also fails, there is a short circuit in the power supply or somewhere in the monitor's other circuits.

Your first step is to determine *where* the problem is. If your monitor uses a separate power supply module, you can often disconnect the supply from the rest of the monitor by removing a cable harness. With the supply isolated, insert another fuse and try the monitor again. Of course, the monitor will not work since the power supply is disconnected, but if the fuse blows once again, you know the fault is somewhere in the power supply module. Turn off and unplug the monitor. You can then check the supply components for a shorted rectifier diode or shorted filter capacitor. You may also check for soldering shorts on the PC board itself. If the new fuse holds, you have a problem outside the supply (in the raster or video amplifier boards). In that case, you may have to replace the raster board. The video amplifier board may also be defective, but that is unlikely.

151

When your power supply circuit is integrated onto the raster board, isolation becomes a problem. Since there is no way to physically separate the power supply from the rest of the monitor, you should start by assuming the fault is in the supply circuit. This is the most likely problem area and is often the easiest to check. Ultimately, you need to turn off and unplug the monitor and then check the power supply circuit for shorted rectifier diodes or shorted filter capacitors. These components are the

Linear power supplies

most frequent cause of supply shorts. You should also keep an eye out for shorted solder connections on the PC board itself. If you find one or more defective components, replace them with *exact* replacement parts. A schematic would certainly help you here, but you can often trace at least most of the supply circuit by eye. If you do not find a defective component, you will have little choice but to replace the entire raster board/power supply assembly.

Symptom 4 *The main ac fuse fails when the power supply is cold.* A monitor draws a fairly large amount of current. During startup, this initial surge (or inrush) of current can stress power supply components. To reduce the effects of inrush current, many power supplies provide one or more current limiting *thermistors,* which are resistors whose resistance changes with temperature. At room temperature, the thermistors exhibit a slightly higher resistance that will limit inrush current and help protect the fuse and other power supply components. Within a few fractions of a second, the power dissipated by the thermistor heats it up. This in turn drops its resistance and allows full current flow into the supply.

The problem is that thermistors eventually break down, and their room-temperature resistance will drop. This negates their effectiveness for inrush protection, and you will see the monitor start snapping fuses occasionally for no apparent reason. In fact, you may mistake this symptom for Symptom 3. The classic sign of inrush problems occurs when a monitor blows a fuse, but works right after you replace the fuse. Turn off and unplug the monitor; then check the values of any thermistors (or posistors) located in the ac filter network (right after the ac power switch). If the measured value is significantly lower than the marked value, replace the thermistor(s). Remember to measure the thermistors cold (otherwise you will see a lower value). If you are unable to locate the thermistor(s) or determine their condition, replace the power supply outright.

Switching power supplies

The great disadvantage to linear power supplies is their tremendous waste. At least half of all power provided to a linear supply is literally thrown away as heat; most of this waste occurs in a regulator. Ideally, if there was "just enough" energy supplied to the regulator to achieve a stable output voltage, regulator waste could be reduced almost entirely.

152

Concepts of switching regulation

Instead of throwing away extra input energy, a *switching* power supply senses the output voltage provided to a load and then switches the power supply circuit on or off as needed to maintain steady levels. A block diagram of a typical switching power supply is shown in Fig. 6-12. There are a variety of configurations that are possible, and Fig. 6-12 illustrates one of these possibilities. You can see the similarities and differences between a switching supply and the linear supply shown in Fig. 6-2.

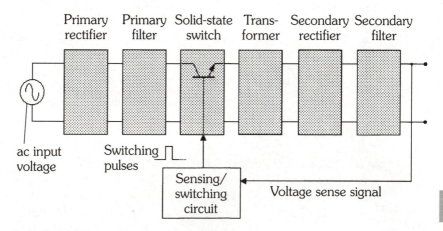

■ **6-12** *Block diagram of a switching power supply.*

Raw ac line voltage entering the supply is immediately converted to pulsating dc and then filtered to provide a "primary dc" voltage. Notice that ac is not transformed before rectification, so primary dc can reach levels approaching 170 V. Remember that ac is 120 volts RMS. Since capacitors charge to the peak voltage (peak = RMS × 1.414), dc levels can be higher than your ac voltmeter readings. High-voltage pulsating dc is as dangerous as ac line voltage and should be treated with extreme caution.

On startup, the switching transistor is turned on and off at a high frequency (usually 20 to 40 kHz) and a long duty cycle. For monitors, the oscillation frequency is typically taken directly from the horizontal raster. The switching transistor breaks up this primary dc into chopped dc which can now be used as the primary signal for a step-down transformer. The duty cycle of chopped dc will affect the ac voltage level generated on the transformer's secondary. A long duty cycle means a larger output voltage (for heavy loads), and a short duty cycle means lower output voltage (for light loads). *Duty cycle* itself refers to the amount of time that a signal is on compared to its overall cycle. Duty cycle is continuously adjusted by the sensing-switching circuit. You can use an oscilloscope to view switching and chopped dc signals.

The ac voltage produced on the transformer's secondary winding (typically a step-down transformer) is *not* a pure sine wave (usually square waves), but it alternates regularly enough to be treated as ac by the remainder of the supply. Secondary voltage is re-rectified and refiltered to form a "secondary dc" voltage that is actually applied to the load. Output voltage is sensed by the sensing-switching circuit which constantly adjusts the chopped dc duty cycle. As load increases on the secondary circuit (more current is drawn by the load), output voltage tends to drop. This is perfectly normal, and the same thing happens in every unregulated supply. However, a sensing circuit detects this voltage drop and increases the switching duty cycle. In turn, the duty cycle for chopped dc increases, which increases the voltage produced by the secondary winding. Output voltage climbs back up again to its desired value—output voltage is regulated.

The reverse will happen as load decreases on the secondary circuit (less current is drawn by the load). A smaller load will tend to make output voltage climb. Again, the same actions happen in an unregulated supply. The sensing-switching circuit detects this increase in voltage and reduces the switching duty cycle. As a result, the duty cycle for chopped dc decreases, and transformer secondary voltage decreases. Output voltage drops back to its desired value—output voltage remains regulated.

Consider the advantages of a switching power circuit. Current is only drawn in the primary circuit when its switching transistor is on, so very little power is wasted in the primary circuit. The secondary circuit will supply just enough power to keep load voltage constant (regulated), but very little power is wasted by the secondary rectifier, filter, or switching circuit. Switching power supplies can reach efficiencies higher than 85 percent (35 percent more efficient than most comparable linear supplies). More efficiency means less heat is generated by the supply, so components can be smaller and packaged more tightly.

Unfortunately, there are several disadvantages to switching supplies that you must be aware of. First, switching supplies tend to act as radio transmitters. Their 20 to 40 kHz operating frequencies can play havoc with radio and television reception, not to mention the circuitry within the monitor itself. This is why you will see most switching supplies somehow covered or shielded in a metal casing. It is critically important that you replace any shielding or ground wires removed during your repair. Strong electromagnetic interference (EMI) can disturb a monitor's operation. Second, the output voltage will always contain some amount of high-frequency ripple. In many applications, this is not enough noise to present interference to the load. In fact, a great many monitors use switching power supplies. Finally, a switching supply often contains more components and is more difficult to troubleshoot than a linear supply. This is often outweighed by the smaller, lighter packaging of switching supplies.

Today, sensing and switching functions can be fabricated right onto an integrated circuit. IC-based switching circuits allow simple, inexpensive circuits to be built as shown in Fig. 6-13. Notice how similar this looks to a linear supply. Alternating current line voltage is transformed (usually stepped down) and then it is rectified and filtered before reaching a switch regulating IC. The IC chops dc voltage at a duty cycle that will provide adequate power to the load. Chopped dc from the switching regulator is filtered by the combination of choke and output filter capacitor to re-form a steady dc signal at the output. The output voltage is sampled back at the IC, which constantly adjusts the chopped dc duty cycle.

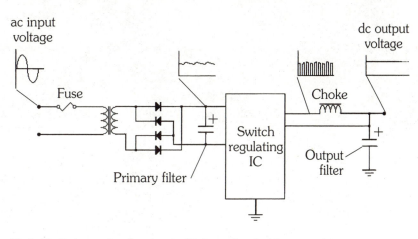

■ **6-13** *Schematic of a simple IC-based switching supply.*

Troubleshooting switching power supplies

Troubleshooting a switching power supply can sometimes be a complex and time-consuming task. Although the operations of rectifier and filter sections are reasonably straightforward, sensing-switching circuits can be complex oscillators that are difficult to follow without a schematic. When symptoms point to a fault in the power supply, it is always acceptable to replace the supply outright. For the purposes of this troubleshooting discussion, consider the transistor-based switching supply shown in the block diagram of Fig. 6-14. The corresponding schematic for the block diagram is shown in Fig. 6-15. Don't let the schematic scare you—it looks more complicated than it actually is.

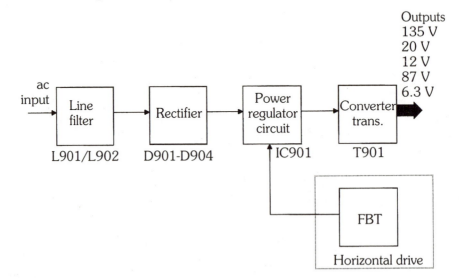

■ **6-14** Block diagram of the Tandy VGM220 power supply. (Courtesy of Tandy Corporation)

The power supply of Fig. 6-15 is a variation of the IC-based switching regulator design. Alternating current is immediately fused through F901 and passed through a series of line filters (L901 and L902 and their associated components). The line filter circuits are used to suppress spikes and surges that may appear occasionally on the ac line. Line-filtered voltage is rectified by a bridge rectifier circuit using diodes D901 to D902. The pulsating dc is filtered and made available to transistor Q901 and transformer T901. Pulses from the flyback transformer (T302) are used to synchronize the switching control IC (IC901) and ultimately fire the switching transistor (Q901). IC901 breaks up the filtered dc into a high-frequency pulse. The pulse frequency matches that of the horizontal sync frequency.

High-frequency pulses feed the transformer T901 at pins 4 and 3. The Q901 circuit driving pins 6, 7, and 1 acts as an "error amplifier" which compensates the transformer's outputs as load conditions change. In turn, the transformer generates a series of voltages which form the power supply outputs. Output pin 10 from T901

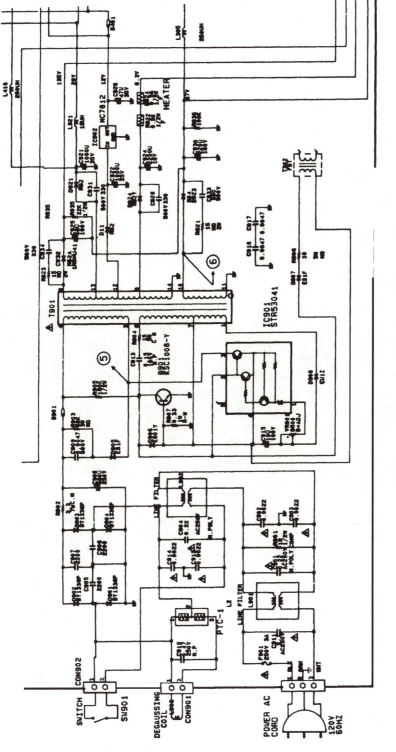

6-15 Schematic fragment of the Tandy VGM220 power supply. (Courtesy of Tandy Corporation)

feeds D923 as a half-wave rectifier whose output is filtered by the 220 μF capacitor (C930) to form the 87 V dc output. Output pin 8 is a low-voltage output that is half-wave rectified by D924 and filtered by the 2200 μF capacitor (C924) to produce a 6.3 V dc supply. This 6.3 V dc powers the CRT cathodes. Output pin 12 supplies the half-wave rectifier D11, which is filtered by the 1000 μF capacitor (C922) to produce the 12 V dc output. To ensure a steady output, regulation is provided by IC902. Output pin 13 supplies the half-wave rectifier D921. The rectifier's output is filtered by the 1000 μF capacitor (C921) to produce a 20 V dc supply. Finally, output pin 9 feeds the half-wave rectifier D932 and is filtered by C925 to generate the 135 V dc supply.

Symptom 1 *The monitor is completely dead. There is no raster and no picture.* When the monitor does not operate and no power indicators are lit, your first suspicion should be the power supply. As with linear supplies, check the ac line voltage entering the supply before beginning any major repair work. Use your multimeter to measure the ac line voltage available at the wall outlet. If you are using an isolation transformer, you should check the transformer's ac output to the supply. Use **extreme** caution whenever measuring ac line voltage levels. Normally, you should read between 105 and 130 V ac to ensure proper supply operation (210 to 240 V ac in Europe). Also see that any voltage selector switch is set appropriately. If you find either very high or low ac voltage, try the monitor in an outlet that provides the correct amount of ac voltage. Unusual line voltage levels may damage your power supply, so proceed cautiously.

If ac line voltage is normal, suspect a main power fuse in the supply. Unplug the line cord and inspect the main fuse. Most monitors provide a panel-mount fuse that you can remove and test. If a panel-mount fuse is not available, you will have to disassemble the monitor to check its fuse(s). You can test a fuse by measuring its continuity. A good fuse should measure as a short circuit (0 Ω), while a failed fuse will measure as an open circuit (infinity). Replace any failed fuse and retest the system. If the fuse continually fails, there is a serious defect elsewhere within the power supply or monitor circuits. Replace the defective fuse(s) *only* with fuses of equal size and current rating.

With the ac supply to the monitor disconnected, check all connectors and wiring to be sure that everything is intact. A loose connector or wire can easily disable the supply. If your supply has failed after being dropped or abused, consider the possibility of printed circuit board damage. Faulty soldering at the factory (or on your test bench) can also cause PC board problems. The power supply PC board may be cracked, one or more component leads may have pulled away from their solder pads, or a trace (or traces) may have broken from impact stress. Examine the PC board assembly carefully.

Apply power and use your multimeter to measure the dc output(s) from the supply. If each output measures correctly, then your trouble lies *outside* of the supply, perhaps in some connector or wiring that provides power to the monitor circuits. A low-output voltage suggests a problem within the supply itself. When symptoms point to the power supply, it is acceptable to replace the entire power supply outright.

When the *entire* power supply is inoperative, use your multimeter to trace ac into the power. As you see in Fig. 6-15, there will be ac present across both line filters (L901 and L901). There should also be ac at the input to the bridge rectifier (the junction of D901 and D902 and the junction of D903 and D904). If ac is not present, there is a circuit interruption somewhere in the line filter circuits. When ac is present across the bridge rectifier, use your oscilloscope to measure the pulse across T901 pins 3 and 4 (the transformer's primary input). You should see pulses at about 290 V. The pulse timing should match the monitor's horizontal sync period (i.e., 31.7 μs for a 31.5 kHz sync frequency). If the pulse is low, absent, or distorted, check the bridge rectifier diodes, D905, and their associated filtering capacitors. A defect here will disable the entire power supply. Check the flyback pulse that drives the switching controller IC901 at pin 2. If this trigger pulse is absent, there may be a defect with the flyback transformer or its rectifier diode D907. If the trigger pulse is present, check the switching output on IC901 pin 3. If the output is absent, replace the switching controller IC. If the output is present, check the error correction circuit of transistor Q901 and its associated components. If everything checks properly but there is still no significant output from T901, one or more of the transformer's primary windings may be defective. Try replacing T901.

When only *one* of the supply outputs is low or absent, check the individual rectifier/filter from the output back to T901. Either the filter capacitor or half-wave rectifier diode for that output may have failed. Use your multimeter to check ac at the anode of the suspect diode. The ac should measure correctly (approximately the same level as you expect the output to be). For example, a secondary ac level of 12 V ac or higher should be expected to support an output of 12 V dc or so. Check dc at the diode-capacitor junction. A low or absent reading probably indicates a faulty diode. Power down the supply and perform a static check of the diode. Replace the diode if it is defective. If the diode is good, perform a static check of the filter capacitor(s). Replace any defective capacitors. If secondary ac voltage is low or absent, there may be a problem with T901. Try replacing T901.

If the problem persists, the trouble may be caused by a more serious circuit fault in the raster or video circuits. Try disconnecting the particular output from the rest of the monitor (easy if the supply is attached to the monitor circuitry by a cable and connector) and measure the "open circuit" output level. If the level returns to a normal value, the fault is likely elsewhere in the monitor. If you see no change in the output, the best course now is simply to replace the power supply.

Symptom 2 *The monitor operates only intermittently. You might see the raster and power indicator LED blink on and off.* Begin by inspecting the ac line voltage into the supply. Be sure that the ac line cord is secured properly at the wall outlet and monitor. Exercise extreme caution whenever measuring ac line voltage. Use your multimeter to measure ac voltage at the power supply's plug. If you are using an isolation transformer as recommended, check the transformer's ac output to the supply. You should normally need 105 to 130 V ac (210 to 240 V ac in Europe). More or less line voltage can cause the supply to malfunction. Inspect every connector or interconnecting wire leading into or out of the power supply. A loose or improperly installed connector can play havoc with supply operation. Pay particular attention to any output connections. A switching power supply must often be connected to its load circuit in order to operate.

In many cases, intermittent operation may be the result of a PC board problem. PC board problems are often the result of physical abuse or impact, but they can also be caused by accidental damage during a repair. The power PC board may be cracked, one or more component leads may have pulled away from their solder pads, or a trace (or traces) may have broken from the stress of impact. Carefully inspect your PC board assembly or replace the power supply outright.

You should also check the supply for *thermal intermittent* problems. Thermal problems typically occur in semiconductor devices such as transistors or ICs. A thermal failure usually occurs when the supply works once it is turned on and then cuts out after some period of operation. A component may work when cool, but fail later on after reaching or exceeding its working temperature. After a monitor quits, check for any unusually hot components. *Never touch an operating circuit with your fingers*—injury is almost certain. Instead, smell around the circuit for any trace of burning semiconductor or try to sense unusually heated air. If you detect an overheated component, spray it with electronics-grade liquid refrigerant available from almost any electronics store. Spray in very short bursts for the best cooling. If normal operation returns, then you have isolated the defective component. Replace any components that behave intermittently. If operation does not return, test any other unusually warm components or replace the entire supply.

Symptom 3 *The main ac fuse fails, and the new replacement fuse also fails.* Remember that a fuse is a protection device. When a fuse opens, it means that too much current has been drawn into the power supply. This can sometimes happen normally due to a power surge on the ac line or a fuse failing from old age. This will shut down the power supply (and monitor) completely, and if you have read Symptom 1 carefully, you will probably find the defective fuse in a matter of minutes. If the failure was just spontaneous, you can replace the fuse, and the monitor will continue to operate normally (though you should still check it thoroughly just to be safe). However, if the replacement fuse also fails, there is a short circuit in the power supply or somewhere in the monitor's other circuits.

Your first step is to determine *where* the problem is. If your monitor uses a separate power supply module, you can often disconnect the supply from the rest of the monitor by removing a cable harness. With the supply isolated, insert another fuse and try the monitor again. Of course, the monitor will not work since the power supply is disconnected, but if the fuse blows once again, you know the fault is somewhere in the power supply module. Turn off and unplug the monitor. You can then check the supply components for a shorted rectifier diode or shorted filter capacitor. The switching components are almost never responsible for this kind of problem. You may also check for soldering shorts on the PC board itself. If the new fuse holds, you have a problem outside the supply (in the raster or video amplifier boards). In that case, you may have to replace the raster board. The video amplifier board may also be defective, but that is unlikely.

When your power supply circuit is integrated onto the raster board, isolation becomes a problem. Since there is no way to physically separate the power supply from the rest of the monitor, you should start by assuming the fault is in the supply circuit. This is the most likely problem area and is often the easiest to check. Ultimately, you need to turn off and unplug the monitor and then check the power supply circuit for shorted rectifier diodes or shorted filter capacitors. These components are the most frequent cause of supply shorts. You should also keep an eye out for shorted solder connections on the PC board itself. If you find one or more defective components, replace them with *exact* replacement parts. A schematic would certainly help you here, especially with a switching power supply, but you may be able to trace at least most of the supply circuit by eye. If you do not find a defective component, you will have little choice but to replace the entire raster board/power supply assembly.

Symptom 4 *The main ac fuse fails when the power supply is cold.* A monitor draws a fairly large amount of current. During startup, this initial surge (or inrush) of current can stress power supply components. To reduce the effects of inrush current, many power supplies provide one or more current limiting *thermistors,* which are resistors whose resistance changes with temperature. At room temperature, the thermistors exhibit a slightly higher resistance that will limit inrush current and help protect the fuse and other power supply components. Within a few fractions of a second, the power dissipated by the thermistor heats it up. This in turn drops its resistance and allows full current flow into the supply.

The problem is that thermistors eventually break down, and their room-temperature resistance will drop. This negates their effectiveness for inrush protection, and you will see the monitor start snapping fuses occasionally for no apparent reason. In fact, you may mistake this symptom for Symptom 3. The classic sign of inrush problems occurs when a monitor blows a fuse, but works right after you replace the fuse. Turn off and unplug the monitor; then check the values of any thermistors (or posistors) located in the ac filter network (right after the ac power switch). If the measured value is significantly lower than the marked value, replace the thermistor(s). Remember to measure the thermistors cold (otherwise you will see a lower value). If you are unable to locate the thermistor(s) or determine their condition, replace the power supply outright.

Backlight power supplies

Liquid crystal displays (LCDs) are visible because light is reflected or transmitted to your eyes. Whatever light emanates from the display is interpreted as being transparent. Light that is absorbed by energized liquid crystal material appears opaque. Light will not always fall evenly or regularly across your small computer's display, so artificial light is used to produce a consistent light source for the display. The artificial lighting used to illuminate LCDs is known as *backlighting*. This part of the chapter covers the power supplies used to drive typical backlights.

Supply requirements and principles

Just about all major backlighting schemes require a high-voltage, low-current signal to power the light source. In many cases, an ac voltage in excess of 200 V is needed. Some larger backlight sources require up to 1000 V dc. To supply a constant ac level, your laptop or notebook computer uses a tiny *inverter* circuit similar to the one shown in Fig. 6-16. Direct current from your laptop's internal battery is fed to an oscillator which "chops" the dc into low-voltage pulsating dc. Pulsating dc is applied across a small, high-ratio step-up transformer which multiplies the pulsating dc into a rough ac signal. This high-voltage ac signal can then

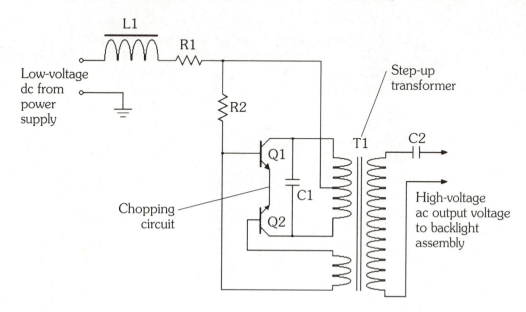

■ **6-16** *Schematic of a typical backlight power inverter circuit.*

be used to run a cold-cathode fluorescent tube (CCFT) or electro-luminescent (EL) panel. Keep in mind that the conversion of dc into ac is virtually opposite to the process used in linear power supplies (thus the term *inverter*) where ac is transformed into dc.

Troubleshooting backlight power supplies

Backlight problems usually manifest themselves in the LCD itself. Without proper lighting, the contrast and brightness of a display will be extremely poor. The display may appear clearly in strong daylight, but may disappear in low light or darkness. When backlight problems occur, you should investigate your inverter supply as well as the particular mechanism (i.e., CCFT or EL panel) producing the light.

Symptom 1 *The backlight appears inoperative. The LCD may seem washed out or invisible in low light.* Remember that virtually all notebook and subnotebook computers are designed to shut down the backlight after some period of inactivity regardless of whether the system is being powered by battery or line voltage. Backlights such as CCFTs and EL panels do not last forever, so disabling the backlight not only saves power during battery operation, but saves the backlight itself. If the backlight cuts out suddenly, it may simply have timed out. Try pressing a key to restore backlight power. You can usually select the backlight timeout period through the system setup software.

Disassemble the display portion of your display to expose the inverter board (typically located behind or next to the LCD). Apply power to the system; then use your multimeter to measure the inverter's dc input voltage. Input voltage usually runs anywhere from 6 to 32 V dc depending on your particular system and backlight type. In any case, you would expect to measure a strong, steady dc voltage. If input voltage is low or absent, an output from your power supply or dc-dc converter may be faulty.

Next, use your multimeter to measure the inverter's ac output voltage. Fluorescent tubes and EL panels typically require from 200 to 1000 V ac for starting and running illumination. If output voltage is low or absent, the inverter circuit is probably defective. You may simply replace the inverter circuit outright or attempt to troubleshoot the inverter to the component level. If output voltage measures an acceptable level, your inverter board is probably working correctly, and the trouble may exist in the light source itself. For example, a CCFT may have failed, or an EL panel may be damaged. Both of these items do wear out due to old age and general wear, so try replacing the suspect light source.

If you elect to try troubleshooting the inverter board itself, you may see from Fig. 6-16 that there is little to fail. Remove all power from the computer and check the oscillator transistors. A faulty transistor can stop your inverter from oscillating, so no ac voltage will be produced. Replace any defective transistors. Beyond faulty transistors, inspect any electrolytic capacitors on the inverter board. A shorted or open tantalum or aluminum electrolytic capacitor may prevent the oscillator from functioning. The transformer may also fail, but transformers are often specialized components that are difficult to find replacements for. If you are unable to locate any obvious component failures, go ahead and replace the inverter board.

High-voltage power supplies

High voltage is perhaps the most critical (and dangerous) aspect of any computer monitor. A CRT requires an extremely high potential to accelerate an electron beam from the cathode to its phosphor-coated face (easily a distance of 12 to 14 in. or more). To accomplish this feat of physics, a positive voltage of 15,000 to 30,000 V dc is applied to the CRT's face through a connection known as the *anode*. It is easy to identify the anode connection—it is underneath the thick red rubber cap on the upper right corner of the CRT (with the neck toward you). Fortunately, the high-voltage system is easier to understand than the conventional power supplies you have seen in this chapter.

A typical high-voltage system is illustrated in the schematic fragment of Fig. 6-17. As you can see, there is really only one critical part: the flyback transformer (FBT) marked T302. The horizontal output transistor (Q302) generates high-current pulses that control the horizontal deflection yoke (H-DY). Horizontal output signals are also fed to the FBT, which boosts the signal to its final high-voltage level. The lower primary winding is connected back to the video circuit and acts as an "error amplifier." This allows the high-voltage level to vary as contrast and brightness are adjusted. The FBT assembly produces three outputs. The high-voltage output is connected directly to the CRT anode through a well-insulated, high-voltage cable. A supplemental output of several hundred volts feeds a small voltage divider network consisting of two potentiometers and a fixed resistor. The top potentiometer controls the higher voltages and is used to drive the CRT's focus electrode(s). The second potentiometer drives the CRT's screen electrode(s).

The voltages generated by an FBT are all *pulsating dc* signals. Since all transformers work with ac rather than dc, there is always

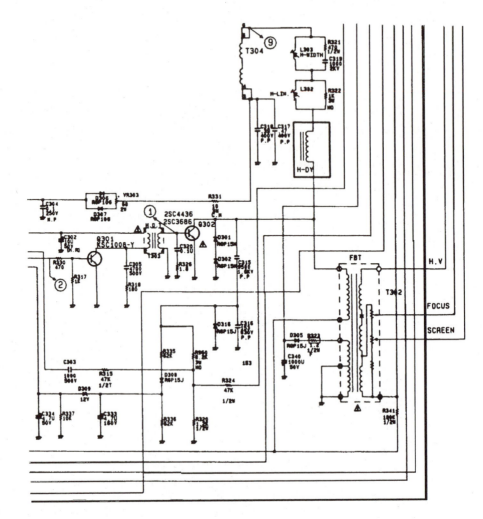

■ 6-17 *Schematic fragment of a high-voltage power supply. (Courtesy of Tandy Corporation)*

a question of how dc is produced by an ac device. The answer is in the small diode located between the top and middle coils of the FBT secondary. This diode forms a half-wave rectifier built right into the FBT. Pulsating high-voltage is smoothed by the characteristic capacitance of the CRT. The pulsating focus and screen voltages are smoothed by filtering components located on the video board attached to the CRT.

Troubleshooting high-voltage power supplies

The loss of high voltage can manifest itself in several ways depending on exactly where the fault occurs, but in virtually all cases, the screen image and raster will be disturbed or disappear completely. Whenever a screen image disappears, you should first suspect a fault in the conventional power supply. By measuring each available output with a multimeter, you can often determine whether the problem is inside or outside of the power supply. When one or more conventional power output levels appear low or absent, concentrate your troubleshooting on the conventional supply. When all outputs appear normal, the problem is likely in the high-voltage system. Refer to Fig. 6-17.

Symptom 1 *There is no raster and no picture, or there is a vertical line against the raster.* The power LED appears steadily lit, and all conventional power supply outputs measure correctly. Make sure that the display contrast and brightness controls are set to acceptable levels. Also check that the video signal cable from the video adapter is connected properly. Use your oscilloscope and measure the horizontal pulse at the collector of the horizontal output transistor (Q302). If the pulse is present, the fault is likely in the FBT assembly. It is possible to test the FBT using the proper test equipment (e.g., a Sencore Monitor Test System), but such equipment is typically too expensive for all but the most committed test benches. Try replacing the FBT. Keep in mind that the FBT is a critical part and must be replaced with an *identical part.* If the problem persists, suspect the CRT. Use a CRT analyzer-rejuvenator (if possible) to check the CRT. If the CRT checks bad or the equipment is not available, try replacing the CRT. Note that the CRT is often the most expensive part of a monitor. Before replacing the CRT, you should carefully weigh the cost of another CRT against the cost of another monitor.

If the horizontal pulse is missing from the collector, use your oscilloscope to check the input at the base of Q302. You should find approximately a 5-V pulse. If the pulse is present, Q302 has failed and should be replaced with an exact replacement part. If the pulse is missing from the base, check the collector of the horizontal switching transistor (Q301). If the pulse is present at the collector of Q301 but missing from the base of Q302, the fault is likely in the horizontal output transformer (T303). Try replacing T303. If you find the pulse signal missing from the collector of Q301, check for the pulse at the base of Q301. If the pulse is present, Q301 is defective and should be replaced. If the pulse is still missing from the base of Q301, the problem is likely in the horizontal output controller (IC301, which is not shown in Fig. 6-17). Try replacing the horizontal controller IC. If you are unable to follow the circuit or locate the defect, you can simply replace the main monitor PC board.

Monochrome monitor troubleshooting

ALTHOUGH MONOCHROME MONITORS DO NOT HAVE THE WIDE consumer appeal of color monitors, the monochrome monitor has secured a permanent place for itself in the PC world (Fig. 7-1). There are many basic PC database, indexing, and record-keeping applications where a color monitor is simply not necessary or appropriate. Their low cost and relative simplicity are likely to keep monochrome monitors in service for many more years. This chapter explains the detailed internal workings of a typical monochrome monitor and illustrates a variety of symptoms and solutions that you can use to repair them. Keep in mind that the component-level discussions in this chapter are based on example circuit fragments. Your specific monitor will likely use different circuitry that might have some similarities.

The monochrome circuits

The best place to start a discussion of monochrome monitors is with a block diagram such as the one in Fig. 7-2. You will probably recognize this diagram from earlier chapters, but this chapter will look at each block down to the component level before discussing troubleshooting. There are three major subsections that comprise the monitor: the video drive circuit, the vertical drive circuit, and the horizontal drive circuit (including the flyback circuit).

Video drive circuit

The video signal driving a monochrome monitor is typically between 1 to 1.5 V. In order to operate the CRT, a video signal must be amplified to about 20 V. As a result, the video drive circuit is primarily a multistage amplifier as shown in Fig. 7-3. The *video amplifier* stage uses Q1 as an input amplifier. An input amplifier boosts the video signal to about 4 V. A simple logic IC (the 7406) acts to condition the video signal—each "on" pixel must be the

■ **7-1** *An NEC MultiSync 5FGp monitor. (NEC Technologies, Inc.)*

same amplitude. This is vital since any noise or variations in the input signal amplitudes will result in brightness fluctuations from pixel to pixel. Each conditioned pixel signal can reach 4.5 V at maximum contrast. A *contrast* control affects the signal amplitude from IC pin 8 to vary the difference between light and dark

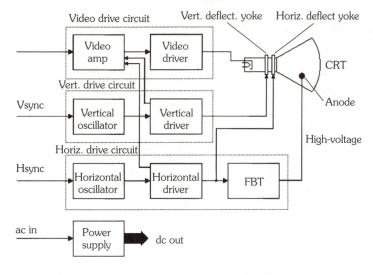

■ **7-2** *Block diagram of a monochrome monitor.*

levels. Conditioned video signals are passed to the final *video driver* stage. Figure 7-3 shows a two-transistor amplifier driver stage. The video driver amplifies logic-level signals to about 20 V. The amplified output is then applied to the CRT.

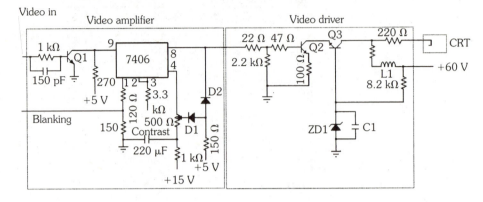

■ **7-3** *Schematic fragment of a monochrome video drive circuit.*

There are a variety of ills that can plague the video drive circuit. When an image disappears entirely (but raster remains), the video signal is not getting through to the monitor, or any one of the video amplifier stages has failed. The fault may also be in the CRT. If there is not even raster, the problem may be in the monitor's power supply or high-voltage system. Short circuits in the CRT focus, screen, or control grids can also disable the display. The troubleshooting section of this chapter covers these types of problems in much more detail.

Vertical drive circuit

A vertical drive circuit is responsible for controlling the vertical position of the raster. The raster sweeps down the display at a constant rate until a vertical synchronization pulse is received. The sync pulse causes the sweep to stop, retrace, and start a new sweep. When a vertical synchronization pulse is applied to the drive circuit shown in Fig. 7-4, the vertical oscillator (or *vertical sawtooth generator*) is forced to fire. Although the unijunction-based oscillator circuit is designed to run at 50 Hz, a faster vertical sync pulse (i.e., 60 or 72 Hz) will "overdrive" the oscillator safely. The ramp signal generated by the unijunction circuit is linearized with feedback provided by Q1. Typical controls allow adjustment of vertical size and vertical linearity.

Unfortunately, the linearized ramp signal at the emitter of Q1 is not nearly powerful enough to drive the vertical deflection yoke di-

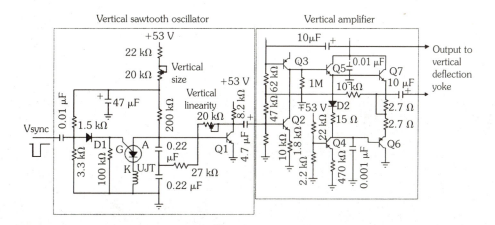

7-4 *Schematic fragment of a vertical drive circuit.*

rectly. A relatively powerful amplifier circuit is required to develop the needed energy. The *vertical amplifier circuit* consists of six transistors arranged in a "push-pull" configuration. Each set of upper and lower transistors provides successively greater signal levels until the final amplifiers (Q7 and Q6) drive the deflection yoke.

Problems that develop in the vertical amplifier will invariably affect the appearance of the CRT image. A catastrophic fault in the vertical oscillator or amplifier will leave a narrow horizontal line in the display. The likeliest cause is the vertical amplifier circuit— both of the final amplifiers may have failed. If only the upper or lower half of an image disappears, only one "side" of the vertical amplifier circuit may have failed. However, any fault that interrupts the vertical sawtooth will disable the vertical deflection entirely. When the vertical deflection is marginal (too expanded or too compressed), suspect a fault in the vertical sawtooth oscillator or its related components. An image that is overexpanded will usually appear "folded over" with a whitish haze along the bottom.

Horizontal drive circuit

The horizontal drive circuit directs the electron beam horizontally across the display. As with vertical drive circuits, the horizontal drive circuit can be divided into two sections: an oscillator and an amplifier. It is the *horizontal oscillator* that generates a carefully shaped square wave (not a sawtooth wave like the vertical oscillator). For the circuit of Fig. 7-5, two IC oscillators (IC1 and IC2) produce the original square wave signal. The horizontal synchronization pulse keeps the oscillator locked at the proper frequency. The rate of scanning for monochrome monitors is usually 15.75 kHz.

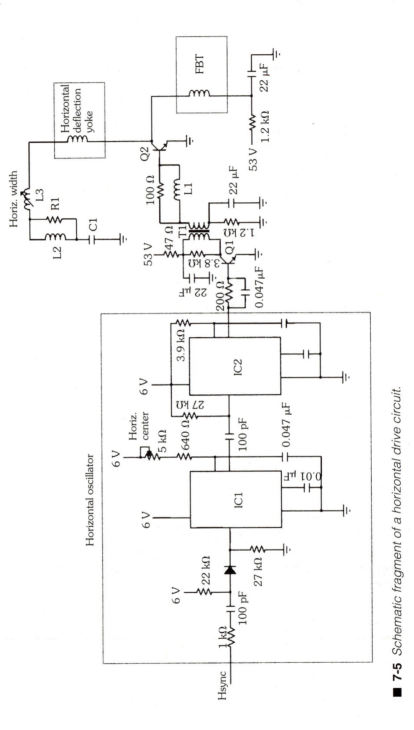

■ **7-5** *Schematic fragment of a horizontal drive circuit.*

By adjusting the offset between the time a sync pulse arrives and the time horizontal pulse starts (with the *horizontal centering* control), it is possible to control centering of the image.

Once the horizontal pulse is generated, it must be amplified in order to drive the horizontal deflection yoke. This is accomplished through a two-stage switch-amplifier consisting of Q1 and Q2. Transistor Q1 is a switch that turns current on and off in the coupling transformer (T1). This current, in turn, drives the *horizontal output transistor* (Q2 or HOT). The HOT (horizontal output transistor) not only drives the horizontal deflection yoke, but also operates the flyback transformer (FBT). When Q2 is turned on, current flows through capacitor C1, resistor R1, inductors L2 and L3, and the horizontal deflection yoke. The oscillator signal effectively draws the electron beam evenly to the right side of the screen. As the beam reaches the right side of the screen, the HOT switches off. The resulting energy discharge draws the electron beam back to the left side of the screen (known as the *retrace*). Horizontal linearity in this circuit is provided by the fixed inductor (L2), and sweep width can be slightly adjusted with the variable inductor (L3).

Problems in the horizontal drive circuit can take several forms. One common manifestation is the loss of horizontal sweep, leaving a vertical line in the center of the display. This is generally due to a fault in the horizontal oscillator circuit rather than the horizontal amplifier circuit. The second common symptom is a loss of image (including raster) and is almost always the result of a failure in the HOT (or high-voltage circuit). Since the HOT also operates the flyback transformer, a loss of horizontal output will disrupt high-voltage generation—the image will disappear.

The flyback circuit

High voltage is a vital element of proper CRT operation. The presence of a large positive potential on the CRT's anode is needed to accelerate an electron beam across the distance between the cathode and CRT phosphor. Electrons must strike the phosphor hard enough to liberate visible light. Under normal circumstances, this requires a potential of 15,000 to 30,000 V. Larger CRTs need higher voltages because there is a greater physical distance to overcome. Monitors generate high-voltage through the *flyback circuit.*

The heart of the high-voltage circuit is the *flyback transformer* (FBT) as shown in Fig. 7-6. The FBT's primary winding is coupled to the horizontal output transistor. Flyback voltage is generated during the horizontal *retrace* (the time between the end of one scan line and the beginning of another) when the sudden drop in deflection signal causes a strong voltage spike on the FBT secondary windings. You will notice that the FBT in Fig. 7-6 provides three secondary windings. Most of the flyback energy is produced in the top winding. This is the high-voltage output. A high-voltage rectifier diode is added to the circuit to form a half-wave rectifier—only positive voltage reaches the CRT anode. The effective capacitance of the CRT anode will act to filter the high-voltage spikes into dc. You can read the high-voltage level with a high-voltage probe as described in Chapter 3.

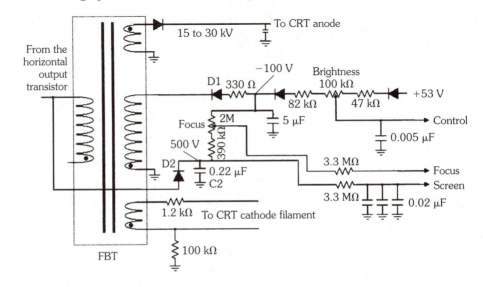

■ **7-6** *Schematic of a CRT grid drive circuit.*

However, the CRT needs additional voltages in order to function. The lower secondary winding on the FBT provides ac for the CRT cathode element. You can tell the cathode by its red glow in the neck of a CRT. The actual ac output from this bottom winding may range from 6.3 to 15 V ac depending on the particular CRT being used. Note that some FBT designs do not provide a cathode ac output. In such an event, the monitor's power supply must provide the ac.

The middle secondary winding produces ac for a −100 V dc supply formed with D1, R1, and C1. Another connection from the HOT collector provides about 500 V dc through the half-wave rectifier D2 and filter capacitor C2. A 53 V dc source is also taken from the monitor's power supply. This network of voltage levels is used to energize the three CRT grids: control, focus, and screen. Once an electron beam is formed from the cathode, the beam passes through the *control grid* first. The control grid affects brightness by limiting the strength of the electron beam. The brightness potentiometer is powered by 53 V dc on one side and −100 V dc on the other. Once brightness is limited, the beam is accelerated by the *screen grid* which is fixed at about 500 V dc. Finally, the beam is narrowed through the *focus grid*. The focus potentiometer is powered by −100 V dc on one side and 500 V dc on the other. Note the large values of resistance used to limit grid current levels. The focused beam is then accelerated to the phosphor screen by anode voltage.

Trouble in the high-voltage circuit can render the monitor inoperative. Typically, a high-voltage fault manifests itself as a loss of image and raster. In many cases where the HOT and deflection signals prove to be intact, the flyback transformer has probably failed causing a loss of output in one or more of the three FBT secondary windings. The troubleshooting procedures in the next section of this chapter will cover high-voltage symptoms and solutions in more detail.

Troubleshooting the monochrome monitor

Any discussion of monitor troubleshooting *must* start with a reminder of the dangers involved. Computer monitors use very high voltages for proper operation. Potentially *lethal* shock hazards exist within the monitor assembly both from ordinary ac line voltage as well as from the CRT anode voltage developed by the flyback transformer. You must exercise **extreme** caution whenever a monitor's outer housings are removed. If you have not yet read about shock hazard dangers and precautions in Chapter 2, *please read and understand that material **now.*** If you are uncomfortable with the idea of working around high voltages, defer your troubleshooting to an experienced technician.

Wrapping it up

When you get your monitor working again and are ready to reassemble it, be very careful to see that all wiring and connectors are routed properly. No wires should be pinched or lodged between the chassis or other metal parts (especially sharp edges). After the wiring is secure, make sure that any insulators, shielding, or protective enclosures are installed. This is even more important for larger monitors with supplemental X-ray shielding. Replace all plastic enclosures and secure them with their full complement of screws.

Postrepair testing and alignment

Regardless of the problem with your monitor or how you go about repairing it, a check of the monitor's alignment is always worthwhile before returning the unit to service. Your first procedure after a repair is complete should be to ensure that the high-voltage level does not exceed the maximum specified value. Recall that excessive high voltage can liberate X radiation from the CRT. Over prolonged exposure, X rays can present a serious biohazard. The high-voltage value is usually marked on the specification plate glued to the outer housing or recorded on a sticker placed somewhere inside the housing. If you cannot find the high-voltage level, refer to service data from the monitor's manufacturer. Once high voltage is correct, you can proceed with other alignment tests. Refer to Chapter 5 for testing and alignment procedures. When testing (and realignment) is complete, it is wise to let the monitor run for 24 hours or so (called a *burn-in test*) before returning it to service. Running the monitor for a prolonged period helps ensure that the original problem has indeed been resolved. This is a form of quality control. If the problem resurfaces, there may be another more serious problem elsewhere in the monitor. The companion disk available for this book contains a burn-in utility that will exercise your monitor.

Symptoms and solutions

Symptom 1 *Raster is present, but there is no image.* When the monitor is properly connected to a PC, a series of text information should appear as the PC initializes. We can use this as our image test. Isolate the monitor by trying an efficiently running monitor on your host PC. If this good monitor works, it indicates that the PC and video adapter are working properly. Reconnect the suspect monitor to the PC and turn up the brightness (and contrast if necessary). You should see a faint white haze covering the display. This is the raster generated by the normal sweep of an electron beam. Remember that the PC *must* be on and running. Without the horizontal and vertical retrace signals provided by the video adapter, there will be no raster.

The video signal is probably being interrupted by a fault in the video drive circuit (refer to Fig. 7-3). Your best course is to use an oscilloscope and check for the presence of video data at the video input. You will probably need to set coupling to ac and then adjust the oscilloscope's time base and triggering to achieve a reasonable trace, which should appear as a series of brief, high-frequency pulses that are all roughly the same amplitude with each pulse corresponding to a pixel that should be lit. Trace the video signal through the circuit. The point at which the signal disappears should be the point of failure. For the example circuit in Fig. 7-3, you would probe the collector of Q1, pin 8 of the IC, the collector of Q2, and the collector of Q3 in that order. Each point should measure larger values until the signal reaches about 20 V at the output of Q3. You will probably need to adjust the oscilloscope's voltage sensitivity to compensate for larger signals as you get closer to the CRT. If, for example, there should be a video signal present at the video input but no signal at the collector of Q1, transistor Q1 (or one of its nearby components) has failed and should be replaced. If you do not have the equipment or inclination to perform this kind of component-level detective work, try replacing the video drive circuit outright. This is usually the small rectangular board attached to the CRT neck.

If the video data check correctly all the way to the CRT (or replacing the video drive circuit does not restore the image), you should suspect a fault in the CRT itself because there is little else that can fail. If you have a CRT tester-rejuvenator available as shown in Chapter 3, you should test the CRT thoroughly for shorted grids or a weak cathode. If the problem cannot be rectified through rejuvenation (or you do not have access to a CRT tester), try replacing the CRT. Keep in mind, however, that a CRT is usually the most expensive part of the monochrome monitor. If each step up to now has not restored your image, you should weigh the economics of replacing the CRT versus scrapping it in favor of a new or rebuilt unit.

Symptom 2 *A single horizontal line appears in the middle of the display.* The horizontal sweep is working properly, but there is no vertical deflection. A fault has almost certainly developed in the vertical drive circuit (refer to Fig. 7-4). Use your oscilloscope to check the sawtooth wave being generated by the vertical oscillator (the collector of Q1). A typical sawtooth wave is illustrated in Fig. 7-7. If the sawtooth wave is missing, the fault is in the oscillator circuit. For the circuit of Fig. 7-4, check the uni-junction transistor (UJT), Q1, D1, and related capacitors. If you are not able to check to the component level, replace the monitor's main PC board.

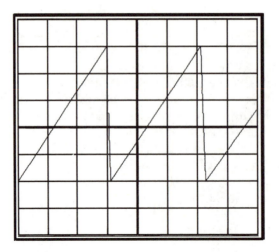

■ 7-7 *Oscilloscope display of a typical saw-tooth pattern.*

If the sawtooth wave is available at the collector of Q1, go immediately to the final output transistors Q6 and Q7. It is not uncommon for one transistor to fail and take the companion transistor with it. Check Q6 and Q7 and replace them if they are defective. If Q6 and Q7 check properly, inspect the previous transistors and their related components. Use the oscilloscope to trace the sawtooth signal until it disappears. The point at which the signal disappears is probably the point of failure. If you do not have the tools or inclination to check the sawtooth signal, replace the monitor's main PC board. Once the image is restored, be sure to check vertical linearity and vertical size as described in Chapter 5. In the event that a new PC board fails to restore vertical deflection, the vertical deflection yoke is probably damaged. This may require replacement of the entire neck assembly and will likely require complete realignment.

Symptom 3 *Only the upper or lower half of an image appears.* The amplifier that drives the vertical deflection yoke is based on a push-pull amplifier design (refer to Fig. 7-4). In many cases, the loss of one transistor in a push-pull pair will result in damage to the associated transistor as well. If only one half of the push-pull team should fail, the image will appear cut in half. This symptom indicates that the vertical sawtooth oscillator is working, but there is an interruption in one or more of the vertical driver transistors. Starting with the final output transistors, check each vertical driver transistor and replace any that are defective. If you do not have the tools or inclination to check and replace devices at the component level, replace the monitor's main PC board. When the image is restored, be sure to check vertical linearity as described in Chapter 5.

Symptom 4 *A single vertical line appears along the middle of the display.* The vertical sweep is working properly, but there is no horizontal deflection. However, in order to even see the display at normal brightness, there must be high voltage present in the monitor—the horizontal drive circuit must be working (refer to Fig. 7-5). The fault probably lies in the horizontal deflection yoke. Check the yoke and all wiring connected to it. It may be necessary to replace the horizontal deflection yoke or the entire yoke assembly.

If horizontal deflection is lost as well as substantial screen brightness, there may be a marginal fault in the horizontal drive circuit. If there is a problem with the horizontal oscillator pulses, the switching characteristics of the horizontal amplifier will change. In turn, this affects high-voltage development and horizontal deflection. Use your oscilloscope to check the square wave generated by the horizontal oscillator as in Fig. 7-5. You should see a wave similar to the one shown in Fig. 7-8. If the wave is distorted, replace the oscillator IC(s). If the horizontal pulse is correct, check the horizontal switching-amplifier transistors (Q1 and Q2). Replace any transistor that appears defective. If the collector signal at the HOT is low or distorted, there may be a short circuit in the flyback transformer primary winding. Try replacing the FBT. If you do not have the tools or inclination to check components to the component level (or the problem persists), replace the monitor's main PC board. When the repair is complete, check the horizontal linearity and size as described in Chapter 5.

Symptom 5 *There is no image and no raster.* When the monitor is properly connected to a PC, a series of text information should appear as the PC initializes. We can use this as our image test. Isolate the monitor by trying an efficiently running monitor on your host PC. If this good monitor works, it indicates that the PC and video adapter are working properly. Reconnect the suspect monitor to the PC and turn up the brightness (and contrast if necessary). Start by checking for the presence of horizontal and vertical synchronization pulses. If pulses are absent, no raster will be generated. If sync pulses are present, there is likely a problem somewhere in the horizontal drive or high-voltage circuits.

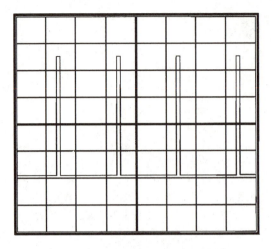

■ **7-8** *Oscilloscope display of a typical horizontal pulse.*

Always suspect a power supply problem and check every output from the supply (especially the 53 V dc output as shown in Fig. 7-5). A low or absent supply voltage will disable the horizontal deflection and high-voltage circuits. If one or more supply outputs are low or absent, you can troubleshoot the power supply circuit as discussed in Chapter 6 or replace the power supply outright (when the power circuit is combined on the monitor's main PC board, the entire main PC board will have to be replaced).

If the supply outputs read correctly, suspect your horizontal drive circuit. Use your oscilloscope to check the horizontal oscillator output at the base of Q1 as shown in Fig. 7-5. You should see a signal similar to the one in Fig. 7-8. If the square wave is low, distorted, or absent, replace the horizontal oscillator IC(s). If a regular pulse is present, the horizontal oscillator is working. Since Q1 is intended to act as a switch, you should also find a pulse at the collector of Q1. If the pulse is severely distorted or absent, Q1 is probably damaged (remove Q1 and test it). If Q1 reads as faulty, it should be replaced. If Q1 reads good, check the horizontal coupling transformer (T1) for shorted or open windings. Try replacing T1 (there is little else that can go wrong in this part of the circuit).

Check the HOT (Q2) next by removing it from the circuit and testing it. If Q2 reads faulty, it should be replaced with an exact replacement part. If Q2 reads good, the fault probably lies in the flyback transformer. Try replacing the FBT. If you do not have the tools or inclination to perform these component-level checks, simply replace the monitor's main PC board outright.

In the event that these steps fail to restore the image, the CRT has probably failed. If you have access to a CRT tester-rejuvenator as discussed in Chapter 3, you can test the CRT. When the CRT measures as bad (and cannot be restored through rejuvenation), it should be replaced. If you do not have a CRT test instrument, you can simply replace the CRT. Keep in mind, however, that a CRT is usually the most expensive part of the monochrome monitor. If each step up to now has not restored your image, you should weigh the economics of replacing the CRT versus scrapping it in favor of a new or rebuilt unit. If you choose to replace the CRT, you should perform a full set of alignments as described in Chapter 5.

Symptom 6 *The image is too compressed or too expanded. A whitish haze may appear along the bottom of the image.* Start by checking your vertical size control to be sure that it was not adjusted accidentally. Since vertical size is a function of the vertical sawtooth oscillator, you should suspect the vertical oscillator circuit. A sawtooth signal that is too large will result in an overexpanded image, whereas a signal that is too small will appear to compress the image. Use your oscilloscope to check the sawtooth signal. For the vertical drive circuit of Fig. 7-4, the signal on the base of Q1 should appear similar to the one shown in Fig. 7-7. If the signal is incorrect, check the UJT and its associated components. Modern monitors often incorporate the oscillator and initial amplifier stages in a single IC. If this is the case, replace the vertical oscillator IC. You may also wish to check the PC board for any cracks or faulty soldering connections around the vertical oscillator circuit. If the problem persists, or you do not have the tools or inclination to perform component-level troubleshooting, simply replace the monitor's main PC board outright.

Symptom 7 *The displayed characters appear to be distorted.* The term *distortion* can be interpreted in many different ways. For our purposes, we will simply say that a distorted image (usually text) is difficult to read. Before even opening your toolbox, check the monitor's location. The presence of stray magnetic fields in close proximity to the monitor can cause bizarre forms of distortion. Try moving the monitor to another location. Remove any electromagnetic or magnetic objects (e.g., motors or refrigerator magnets) from the area. If the problem persists, the monitor is likely at fault.

If only certain areas of the display appear affected (or affected worse than other areas), the trouble is probably due to poor linearity (either horizontal, vertical, or both). If raster speed varies across the display, the pixels in some areas of the image may appear too close together, while the pixels in other areas of the image may appear too far apart. You can check and correct horizontal and vertical linearity using the procedures found in Chapter 5. Alignment test patterns are available on the companion disk for this book. If alignment fails to correct poor linearity, your best course is often simply to replace the monitor's main PC board.

If the image is difficult to read because it is out of focus, you should check the focus alignment using the procedure outlined in Chapter 5. If you cannot achieve a sharp focus using controls either on the front panel of the monitor or on the flyback transformer assembly, there is probably a fault in the flyback transformer. As you see in the example circuit of Fig. 7-6, several voltages are needed to drive the CRT grids. If you also notice problems with screen brightness and brightness control, the −100-V dc supply may be failing. Try replacing the FBT. If the problem persists, your best course is often simply to replace the monitor's main PC board.

Symptom 8 *The display appears wavy.* There are visible waves appearing along the edges of the display as the image sways back and forth. This is almost always the result of a power supply problem— one or more outputs are failing. Use your multimeter and check each supply output. If you find a low or absent output, you can proceed to troubleshoot the supply as discussed in Chapter 6 or you can simply replace the supply outright. If the power supply is integrated onto the main PC board, you will have to replace the entire main PC board.

Symptom 9 *The display is too bright or too dim.* Before opening the monitor, be sure to check the brightness and contrast controls. If the controls had been accidentally adjusted, set contrast to maximum and adjust the brightness level until a clear, crisp display is produced. When front-panel controls fail to provide the proper display (but focus seems steady), suspect a fault in the monitor's power supply. Refer to the example schematic of Fig. 7-6. If the 53-V dc supply is too low or too high, brightness levels controlling the CRT screen grid will shift. If you find one or more low outputs from the power supply, you can troubleshoot the supply as discussed in Chapter 6 or replace the supply outright. For those monitors that incorporate the power supply on the main PC board, the entire main PC board will have to be replaced. If both brightness and focus seem out, there is still probably a voltage problem, and you should refer to Symptom 7.

Symptom 10 *You see visible raster scan lines in the display.* The very first thing you should do is check the front-panel brightness and contrast controls. If contrast is set too low and/or brightness is set too high, raster will be visible on top of the image. This will tend to make the image appear a bit fuzzy. If the front-panel controls cannot eliminate visible raster from the image, chances are that you have a problem with the power supply. Use your multimeter and check each output from the supply. If one or more outputs appear too high (or too low), you can troubleshoot the supply as described in Chapter 6 or replace the supply outright. If the supply is integrated with the monitor's main PC board, the entire PC board will have to be replaced.

If focus is also affected when raster appears, you should check the CRT control voltages shown in the example schematic of Fig. 7-6. A marginal fault in the FBT or horizontal output circuit can result in fluctuations in the −100- or 500-V dc sources. A shift in either of these sources will adversely change the brightness level (−100 V dc) or screen grid level (500 V dc) as well as focus. If the −100-V dc source is too high or too low, try replacing the FBT. If the 500-V dc source is too high or too low, you may also try replacing the FBT or troubleshoot the related components in the horizontal deflection yoke as shown in the example schematic of Fig. 7-5. If you do not have the tools or inclination to tackle component-level troubleshooting, feel free to replace the main PC board outright.

Symptom 11 *The display flickers or cuts out when the video cable is moved.* Check the video cable's connection to the video adapter at the PC. A loose connection will almost certainly result in such intermittent problems. If the connection is secure, there is an intermittent connection in the video cable. Before replacing the cable, check its connections within the monitor itself. When connections are intact, replace the intermittent video cable outright. Do not bother cutting or splicing the cable—any breaks in the signal shielding will allow signal noise to introduce distortion.

Symptom 12 *The image expands in the horizontal direction when the monitor gets warm.* One or more components in the horizontal retrace circuit are weak and are slightly changing value once the monitor gets warm. Turn off and unplug the monitor. You should inspect any capacitors located around the horizontal output transistor (HOT). Typical suspect components would include the 22 mF capacitors around Q1 and Q2 (from Fig. 7-5). The problem is that thermal problems such as this can be extremely difficult to isolate because you can't measure capacitor values while the monitor is running, and after the monitor is turned off, the parts will usually cool too quickly to catch a thermal problem. It is often most effective simply to replace several of the key capacitors around the HOT outright. If you don't want to bother with individual components, replace the raster board.

Symptom 13 *The image shrinks in the horizontal direction when the monitor gets warm.* This is another thermal-related problem which indicates either a weakness in one or more components or a mild soldering-related problem. Turn off and unplug the monitor. Start by checking for a poor solder connection, especially around the horizontal deflection yoke wiring, the horizontal output transistor (HOT), and the FBT. If nothing appears obvious, you might consider resoldering all of the components in the HOT area of the raster board.

If problems continue, suspect a failure in the HOT itself. Semiconductors rarely become marginal; they either work or they don't. Still, semiconductor junctions can become unstable when temperatures change and result in circuit characteristic changes. You could also try replacing the HOT outright.

It is also possible that one or more midrange power supply outputs (i.e., 12 or 53 V) are sagging when the monitor warms up. Use a voltmeter and measure the outputs from your power supply. If the 12- or 53-V outputs appear to drop slightly once the monitor has been running for a bit, you should troubleshoot the power supply as described in Chapter 6.

Symptom 14 *High voltage fails after the monitor is warm.* There are a large number of possible causes behind this problem, but no matter what permutation you find, you will likely be dealing with soldering problems or thermal-related failures. Turn off and unplug the monitor. Inspect the HOT's heat sink assembly. There may be a bad solder connection on the heat sink ground. There may also be an open solder connection on one or more of the FBT pins. If you cannot locate a faulty soldering connection, you may simply choose to resolder all of the connections in the flyback area.

If the problem persists, you should suspect that either your HOT or FBT is failing under load (after the monitor warms up). One possible means of isolating the problem is to measure pulses from the HOT output with your oscilloscope. If the pulses stop at the same time your high voltage fails, you can suspect a problem with your HOT or other horizontal components. Try replacing the HOT. If high voltage fails but the HOT pulses remain, your FBT has probably failed. Replace the FBT. If you do not have an oscilloscope, try replacing the HOT first because that is the least expensive part; then replace the FBT if necessary.

In the unlikely event that both a new HOT and FBT do not correct the problem, you should carefully inspect the capacitors in the HOT circuit. One or more might be failing. Unfortunately, it is very difficult to identify a marginal capacitor (especially one that is suffering from a thermal failure). You may try replacing the major capacitors in the HOT circuit or replace the raster board entirely.

Symptom 15 *The image blooms intermittently.* The amount of high voltage driving the CRT is varying intermittently. Since high voltage is related to the HOT circuit and FBT, you should concentrate your search in those two areas of the raster board. Examine the soldering of your HOT and FBT connections, especially the ground connections if you can identify them. You may try resoldering all of the connections in those areas (remember to turn off and unplug the monitor before soldering). There may also be a ground problem on the video amplifier board which allows all three color signals to vary in amplitude. When this happens, the overall brightness of the image changes, and the image may grow or shrink a bit in response. Try resoldering connections on the video amplifier board.

If the problem remains (even after soldering), your FBT may be failing, probably due to an age-related internal short. High-end test equipment such as Sencore's monitor test station provides the instrumentation to test an FBT. If you do not have access to such dedicated test equipment, however, try replacing the FBT assembly. If you do not have the time or inclination to deal with component replacement, go ahead and replace the raster board outright. In the unlikely event that your problem persists, suspect a fault in the CRT itself. If you have access to a CRT tester-rejuvenator, you can check the CRT's operation. Some weaknesses in the CRT may be corrected (at least temporarily) by rejuvenation. If the fault cannot be corrected, you may have to replace the CRT.

Symptom 16 *The image appears out of focus.* Before suspecting a component failure, try adjusting the focus control. In most cases, the focus control is located at the rear housing or adjacent to the FBT. Keep in mind that the focus control should be adjusted with brightness and contrast set to optimum values—excessively bright images may lose focus naturally. If the focus control is unable to restore a proper image, check the CRT focus voltage. In Fig. 7-6, you can find the focus voltage off an FBT tap. If the focus voltage is low (often combined with a dim image), you may have a failing FBT. It is possible to test the FBT if you have the specialized test instrumentation; otherwise, you should just replace the FBT outright. If you lack the time or inclination to replace the FBT, you can simply replace the raster board.

If a new FBT does not resolve your focus problem, suspect a fault in the CRT, probably in the focus grid. You can use a CRT tester-rejuvenator to examine the CRT, and it may be possible to restore normal operation (at least temporarily). If you do not have such equipment, you will simply have to try a new CRT.

Symptom 17 *The image appears to flip or scroll horizontally.* There is a synchronization problem in your horizontal raster circuit. Begin by checking the video cable to be sure that it is installed and connected securely. Cables that behave intermittently (or that appear frayed or nicked) should be replaced. If the cable is intact, suspect a problem in your horizontal circuit. If there is a horizontal sync (or "horizontal hold") adjustment on the raster board, adjust it in small increments until the image snaps back into sync. If there is no such adjustment on your particular monitor, try resoldering all of the connections in the horizontal processing circuit. If the problem persists, replace the horizontal oscillator IC or replace the entire raster board.

Symptom 18 *The image appears to flip or scroll vertically.* There is a synchronization problem in your vertical raster circuit. Begin by checking the video cable to be sure that it is installed and connected securely. Cables that behave intermittently (or that appear frayed or nicked) should be replaced. If the cable is intact, suspect a problem in your vertical circuit. If there is a vertical sync (or "vertical hold") adjustment on the raster board, adjust it in small increments until the image snaps back into sync. If there is no such adjustment on your particular monitor, try resoldering all of the connections in the vertical processing circuit. If the problem persists, replace the vertical oscillator IC or replace the entire raster board.

Symptom 19 *The image appears to shake or oscillate in size.* This may occur in bursts, but it typically occurs constantly. In most cases, it is due to a fault in the power supply, usually the 53-V (B+) output. Try measuring your power supply outputs with an oscilloscope and see if an output is varying along with the screen size changes. If you locate such an output, the filtering portion of that output may be malfunctioning. Track the output back into the supply and replace any defective components. If you are unable to isolate a faulty component, replace the power supply. When the power supply is integrated onto the raster board, you may have to replace the raster board entirely.

If the outputs from your power supply appear stable, you should suspect a weak capacitor in your horizontal circuit. Try resoldering the FBT, HOT, and other horizontal circuit components to eliminate the possibility of a soldering problem. If the problem remains, you will have to systematically replace the capacitors in the horizontal circuit. If you do not have the time or inclination to replace individual components, replace the raster board outright.

8

Color monitor troubleshooting

COLOR MONITORS HAVE COME A LONG WAY IN THE LAST decade, and they now constitute the vast majority of monitors now in service. The bulky, four-color CGA monitors of the early 1980s have long since been replaced by high-resolution displays, and most current models are capable of reproducing virtually photographic-quality images. The popularity of color goes far beyond the realm of simply creating "attractive" images. Color monitors (Fig. 8-1) have helped to open up entire industries such as computer-aided design and animation, desktop publishing, and computerized photo processing. This chapter explains the detailed internal workings of a typical color monitor and illustrates a variety of symptoms and solutions that you can use to repair them. Keep in mind that the component-level discussions in this chapter are based on example circuit fragments. Your specific monitor will likely use different circuitry that might have some similarities.

The color circuits

In order to have a full understanding of color monitors, it is best to start with a block diagram. The block diagram for a VGA monitor is shown in Fig. 8-2. The first thing you may notice about the color monitor is just how closely it resembles a monochrome CRT. The raster circuits are virtually identical. The improvements needed to form a color system are in the video drive circuits and the CRT. Three complete video drive circuits are needed (one for each primary color). While early color monitors used logic levels to represent video signals, current monitors use *analog* signals which allow the intensity of each color to be varied. The CRT is designed to provide three electron beams that are directed at corresponding color phosphors. By varying the intensity of each electron beam, virtually any color can be produced. For all practical pur-

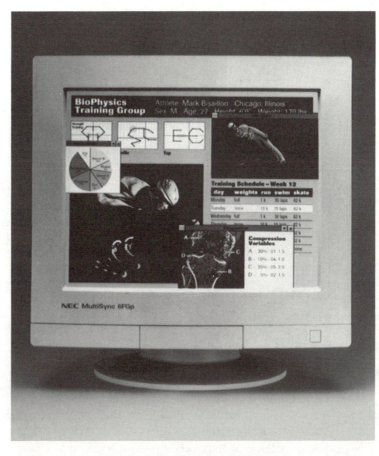

■ **8-1** *An NEC MultiSync 6FGp monitor. (NEC Technologies, Inc.)*

poses, the color monitor can be considered in three subsections: the video drive circuits, the vertical drive circuit, and the horizontal drive circuit (including the high-voltage system).

Video drive circuits

The schematic diagram for a typical red, green, and blue (RGB) drive circuit is shown in Fig. 8-3. This schematic is actually part of a Tandy VGM220 analog color monitor. You will see that there are three separate video drive circuits. Components with a 5xx designation (e.g., IC501) are part of the red video drive circuit. The 6xx designation (e.g., Q602) shows a part in the green video drive circuit. A 7xx marking (e.g., C704) indicates a component in the blue video drive circuit. Other components marked with 8xx designations (e.g., Q803) are included to operate the CRT control grid. Let's walk through the operation of one of these video circuits.

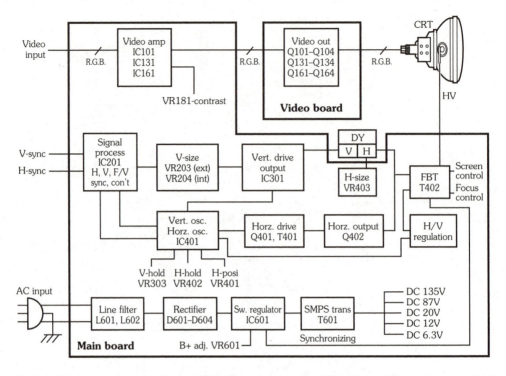

■ **8-2** *Block diagram of a Tandy VGM340 color monitor. (Courtesy of Tandy Corporation)*

The red analog signal is filtered by the small array of F501. The ferrite beads on either side of the small filter capacitor serve to reduce noise that may otherwise interfere with the weak analog signal. The video signal is amplified by transistor Q501. Potentiometer VR501 adjusts the signal *gain* (the amount of amplification applied to the video signal). Collector signals are then passed to the differential amplifier circuit in IC501. Once again, noise is a major concern in color signals, and differential amplifiers help to improve signal strength while eliminating noise. The resulting video signal is applied to a push-pull amplifier circuit consisting of Q503 and Q504 and then fed to a subsequent push-pull amplifier pair of Q505 and Q506. Potentiometer VR502 controls the amount of dc bias used to generate the final output signal. The output from this final amplifier stage is coupled directly to the corresponding CRT video control grid. The remaining two drive circuits both work the same way.

Problems with the video circuits in color monitors rarely disable the image entirely. Even if one video drive circuit should fail, there are still two others to drive the CRT. Of course, the loss of one primary color will severely distort the image colors, but the image

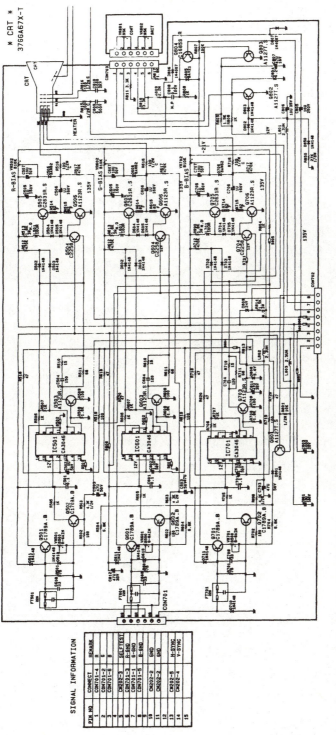

■ 8-3 Schematic of a Tandy VGM220 video drive circuit. (Courtesy of Tandy Corporation)

should still be visible. You can tell when one of the video drive circuits fails; the faulty circuit will either saturate the display with that color or cut that color out completely. For example, if the red video drive circuit should fail, the resulting screen image will either be saturated with red or red will be absent (leaving a greenish-blue or cyan image).

Vertical drive circuit

The vertical drive circuit is designed to operate the monitor's vertical deflection yoke (dubbed V-DY). To give you a broad perspective of vertical drive operation and its interrelation to other important monitor circuits, Fig. 8-4 illustrates the vertical drive, horizontal drive, high voltage, and power supply circuits all combined in the same schematic. This schematic is essentially the main PC board (the raster board) for the Tandy VGM220 monitor. Components marked with 4xx numbers (e.g., IC401) are part of the vertical drive system.

The vertical sync pulses enter the monitor at connector CH202 (the line marked *V*). A simple exclusive-OR gate (IC201) is used to condition the sync pulses and select the video mode being used. Since the polarity of horizontal and vertical sync pulses will be different for each video mode, IC201 detects those polarities and causes the digitally controlled analog switch (IC401) to select one of three vertical size (V-SIZE) control sets which is connected to the vertical sawtooth oscillator (IC402). This mode-switching circuit allows the monitor to autosize the display.

The vertical sync pulse fires the vertical sawtooth oscillator on pin 2 of IC402. The frequency of the vertical sweep is set to 60 Hz, but can be optimized by adjusting the vertical frequency control (V-FREQ) VR404. It is highly recommended that you do *not* attempt to adjust the vertical frequency unless you have an oscilloscope available. Vertical linearity (V-LIN) is adjusted through potentiometer VR405. Vertical centering (V-CENTER) is controlled through VR406. As you recall from in Chapter 5, linearity and centering adjustments should only be made while displaying an appropriate test pattern. It is interesting to note that there are no discrete power amplifiers needed to drive the vertical deflection yoke—IC402 pin 6 drives the deflection yoke directly through an internal power amplifier.

The pincushion circuit forms a link between the vertical and horizontal deflection systems through the pincushion transformer (T304). Transistors Q401 and Q402 form a compensator circuit that slightly modulates horizontal deflection. This prevents distortion in the image when projecting a flat, two-dimensional image

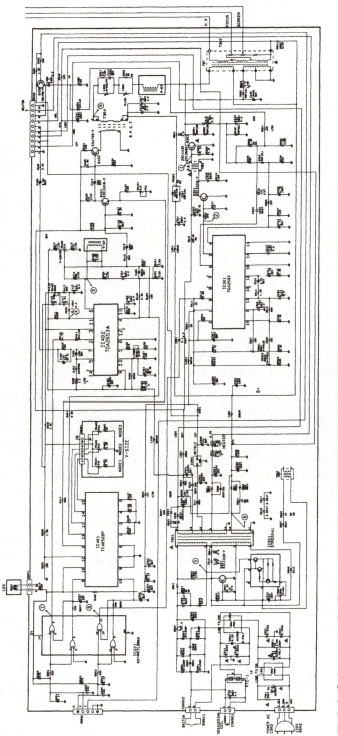

■ 8-4 *Schematic of a Tandy VGM220 raster/high-voltage circuit. (Courtesy of Tandy Corporation)*

The color circuits

onto the CRT's curved surface. Potentiometer VR407 provides the pincushion control (PCC). As with other alignments, you should not attempt to adjust the pincushion unless an appropriate test pattern is displayed.

Problems that develop in the vertical amplifier will invariably affect the appearance of the CRT image. A catastrophic fault in the vertical oscillator or amplifier will leave a narrow horizontal line in the display. The likeliest cause is the vertical drive IC (IC402) since that component handles both sawtooth generation and amplification. If only the upper or lower half of an image disappears, only one part of the vertical amplifier in IC402 may have failed. However, any fault on the PC board that interrupts the vertical sawtooth will disable vertical deflection entirely. When the vertical deflection is marginal (too expanded or too compressed), suspect a fault in IC402, but its related components may also be breaking down. An image that is overexpanded will usually appear "folded over" with a whitish haze along the bottom. It is interesting to note that vertical drive problems do not affect display colors.

Horizontal drive circuit

The horizontal drive circuit is responsible for operating the horizontal deflection yoke (H-DY). It is this circuit that sweeps the electron beams left and right across the display. To understand how the horizontal drive works, you should again refer to the schematic of Fig. 8-4. All components marked with 3xx numbers (e.g., IC301) relate to the horizontal drive circuit. Horizontal sync signals enter the monitor at connector CH202 (the line marked H) and are conditioned by the executive-OR gates of IC201. Conditioned sync pulses fire the horizontal oscillator (IC301). Horizontal frequency should be locked at 31.5 kHz, but potentiometer VR302 can be used to optimize the frequency. Do not attempt to adjust horizontal frequency unless you have an oscilloscope available. Horizontal phase can be adjusted with VR301. You should avoid altering any alignments until a suitable test pattern is displayed as discussed in Chapter 5.

IC301 is a highly integrated device which is designed to provide precision horizontal square wave pulses to the driver transistors Q301 and Q302. IC301 pin 3 provides the horizontal pulses to Q301. Transistor Q301 switches on and off, causing current pulses in the horizontal output transformer (T303). Current pulses produced by the secondary winding of T303 fire the horizontal output transistor (Q302). Output from the HOT drives the horizontal deflection yoke (H-DY). The deflection circuit includes two adjustable coils to control horizontal linearity (H-LIN; L302) and horizontal width (H-WIDTH; L303). You will also notice that the collector signal from Q302 is directly connected to the flyback transformer. Operation of the high-voltage system is covered in the next section.

Problems in the horizontal drive circuit can take several forms. One common manifestation is the loss of horizontal sweep leaving a vertical line in the center of the display. This is generally due to a fault in the horizontal oscillator (IC301) rather than the horizontal driver transistors. The second common symptom is a loss of image (including raster) and is almost always the result of a failure in the HOT (or high-voltage circuit). Since the HOT also operates the flyback transformer, a loss of horizontal output will disrupt high-voltage generation—the image will disappear.

The flyback circuit

The presence of a large positive potential on the CRT's anode is needed to accelerate an electron beam across the distance between the cathode and CRT phosphor. Electrons must strike the phosphor hard enough to liberate visible light. Under normal circumstances, this requires a potential of 15,000 to 30,000 V. Larger CRTs need higher voltages because there is a greater physical distance to overcome. Monitors generate high voltage through the *flyback circuit.*

The heart of the high-voltage circuit is the *flyback transformer* (FBT) as shown in Fig. 8-4. The FBT's primary winding is directly coupled to the horizontal output transistor (Q302). Another primary winding is used to compensate the high-voltage level for changes in brightness and contrast. Flyback voltage is generated during the horizontal *retrace* (the time between the end of one scan line and the beginning of another) when the sudden drop in deflection signal causes a strong voltage spike on the FBT secondary windings. You will notice that the FBT in Fig. 8-4 provides one multitapped secondary winding. The top-most tap from the FBT secondary provides high voltage to the CRT anode. A high-voltage rectifier diode added to the FBT assembly forms a half-wave rectifier; only positive voltages reach the CRT anode. The effective capacitance of the CRT anode will act to filter the high-voltage spikes into dc. You can read the high-voltage level with a high-voltage probe as described in Chapter 3. The CRT needs additional voltages in order to function. The lower tap from the FBT secondary supplies voltage to the focus and screen grid adjustments. These adjustments, in turn, drive the CRT directly.

Trouble in the high-voltage circuit can render the monitor inoperative. Typically, a high-voltage fault manifests itself as a loss of image and raster. In many cases where the HOT and deflection signals prove to be intact, the FBT has probably failed, causing a loss of output in one or more of the three FBT secondary windings. The troubleshooting procedures in the next section of this chapter will cover high-voltage symptoms and solutions in more detail.

Construction

Before jumping right into troubleshooting, it would be helpful to understand how the circuits shown in Fig. 8-4 are assembled. A wiring diagram for the Tandy VGM220 is shown in Fig. 8-5. There are two PC boards: the video drive PC board and the main PC board. The main PC board contains the raster circuits, power supply, and high-voltage circuitry. The video drive PC board contains red, green, and blue video circuits. Video signals, focus grid voltage, screen grid voltage, and brightness and contrast controls connect to the video drive board. The video PC board plugs into the CRT at its neck (the diagram of Fig. 8-5 shows a large round connector which plugs into the CRT). A power switch, power LED, and CRT degaussing coil plug into the main PC board. There are also connections at the main PC board for the ac line cord and video sync signals.

Troubleshooting the color monitor

Any discussion of monitor troubleshooting *must* start with a reminder of the dangers involved. Computer monitors use very high voltages for proper operation. Potentially *lethal* shock hazards exist within the monitor assembly both from ordinary ac line voltage as well as from the CRT anode voltage developed by the flyback transformer. You must exercise **extreme** caution whenever a monitor's outer housings are removed. If you have not yet read about shock hazard dangers and precautions in Chapter 2, *please read and understand that material **now.*** If you are uncomfortable with the idea of working around high voltages, defer your troubleshooting to an experienced technician.

Wrapping it up

When you get your monitor working again and are ready to reassemble it, be very careful to see that all wiring and connectors are routed properly. No wires should be pinched or lodged between the chassis or other metal parts (especially sharp edges). After the wiring is secure, make sure that any insulators, shielding, or protec-

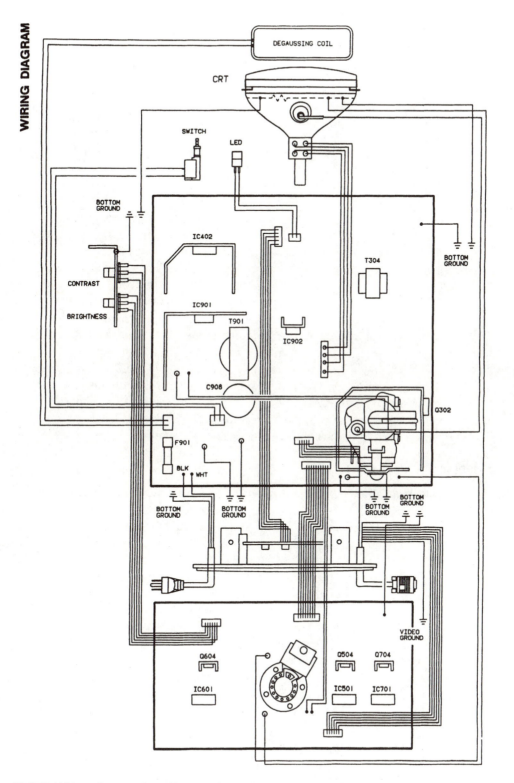

■ **8-5** *Wiring diagram for a Tandy VGM220 monitor. (Courtesy of Tandy Corporation)*

tive enclosures are installed. This is even more important for larger monitors with supplemental X-ray shielding. Replace all plastic enclosures and secure them with their full complement of screws.

Postrepair testing and alignment

Regardless of the problem with your monitor or how you go about repairing it, a check of the monitor's alignment is always worthwhile before returning the unit to service. Your first procedure after a repair is complete should be to ensure that the high-voltage level does not exceed the maximum specified value. Recall that excessive high voltage can liberate X radiation from the CRT. Over prolonged exposure, X rays can present a serious biohazard. The high-voltage value is usually marked on the specification plate glued to the outer housing or recorded on a sticker placed somewhere inside the housing. If you cannot find the high-voltage level, refer to service data from the monitor's manufacturer. Once high voltage is correct, you can proceed with other alignment tests. Refer to Chapter 5 for testing and alignment procedures. When testing (and realignment) is complete, it is wise to let the monitor run for 24 hours or so (called a *burn-in test*) before returning it to service. Running the monitor for a prolonged period helps ensure that the original problem has indeed been resolved. This is a form of quality control. If the problem resurfaces, there may be another more serious problem elsewhere in the monitor. The companion disk available for this book also contains a burn-in feature that will exercise your monitor.

Symptoms and solutions

Symptom 1 *The image is saturated with red or appears greenish-blue (cyan).* If there are any user color controls available from the front or rear housings, make sure those controls have not been accidentally adjusted. If color controls are set properly (or not available externally), the red video drive circuit has probably failed. Refer to the example circuit of Fig. 8-6. Use your oscilloscope to trace the video signal from its initial input (at CN101) to the final output at the CRT. If there is no red video signal at the amplifier input (i.e., pin 9 of IC101), check the connection between the monitor and the video adapter board. If the connection is intact, try an efficiently running monitor. If the problem persists on this good monitor, replace the computer's video adapter board. As you trace the video signal, you can compare the signal to characteristics at the corresponding points in the green or blue video circuits. The point at which the signal disappears is probably the point of failure, and the offending component should be replaced. If you do not

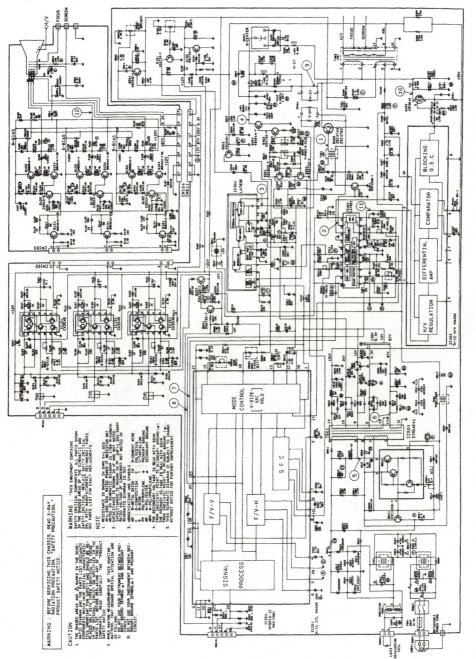

8-6 Schematic of a Tandy VGM340 monitor. (Courtesy of Tandy Corporation)

have the tools or inclination to perform component-level troubleshooting, try replacing the video amplifier PC board entirely.

If the video signal measures properly all the way to the CRT (or a new video drive PC board does not correct the problem), suspect a fault in the CRT itself. The corresponding cathode or video control grid may have failed. If you have access to a CRT tester-rejuvenator, test the CRT. If the CRT measures bad (and cannot be recovered through any available rejuvenation procedure), it should be replaced. Keep in mind that a color CRT is usually the most expensive component in the monitor. As with any CRT replacement, you should carefully consider the economics of the repair versus buying a new or rebuilt monitor.

Symptom 2 *The image is saturated with blue or appears yellow.* If there are any user color controls available from the front or rear housings, make sure those controls have not been accidentally adjusted. If color controls are set properly (or not available externally), the blue video drive circuit has probably failed. Refer to the example circuit of Fig. 8-6. Use your oscilloscope to trace the video signal from its initial input to the final output. If there is no blue video signal at the amplifier input (i.e., pin 9 of IC161), check the connection between the monitor and the video adapter board. If the connection is intact, try an efficiently running monitor. If the problem persists on this good monitor, replace the computer's video adapter board. As you trace the video signal, you can compare the signal to characteristics at the corresponding points in the green or red video circuits. The point at which the signal disappears is probably the point of failure, and the offending component should be replaced. If you do not have the tools or inclination to perform component-level troubleshooting, try replacing the video drive PC board entirely.

If the video signal measures properly all the way to the CRT (or a new video drive PC board does not correct the problem), suspect a fault in the CRT itself. The corresponding cathode or video control grid may have failed. If you have access to a CRT tester-rejuvenator, test the CRT. If the CRT measures bad (and cannot be recovered through any available rejuvenation procedure), it should be replaced. Keep in mind that a color CRT is usually the most expensive component in the monitor. As with any CRT replacement, you should carefully consider the economics of the repair versus buying a new or rebuilt monitor.

Symptom 3 *The image is saturated with green or appears bluish-red (magenta).* If there are any user color controls available from the front or rear housings, make sure those controls have not been accidentally adjusted. If color controls are set properly (or not available externally), the green video drive circuit has probably failed. Refer to the example circuit of Fig. 8-6. Use your oscilloscope to trace the video signal from its initial input to the final output. If there is no green video signal at the amplifier input (i.e., pin 9 of IC131), check the connection between the monitor and the video adapter board. If the connection is intact, try an efficiently running monitor. If the problem persists on this good monitor, replace the video adapter board. As you trace the video signal, you can compare the signal to characteristics at the corresponding points in the red or blue video circuits. The point at which the signal disappears is probably the point of failure, and the offending component should be replaced. If you do not have the tools or inclination to perform component-level troubleshooting, try replacing the video drive PC board entirely.

If the video signal measures properly all the way to the CRT (or a new video drive PC board does not correct the problem), suspect a fault in the CRT itself. The corresponding cathode or video control grid may have failed. If you have access to a CRT tester-rejuvenator, test the CRT. If the CRT measures bad (and cannot be recovered through any available rejuvenation procedure), it should be replaced. Keep in mind that a color CRT is usually the most expensive component in the monitor. As with any CRT replacement, you should carefully consider the economics of the repair versus buying a new or rebuilt monitor.

Symptom 4 *Raster is present, but there is no image.* When the monitor is properly connected to a PC, a series of text information should appear as the PC initializes. We can use this as our baseline image. Isolate the monitor by trying an efficiently running monitor on your host PC. If this good monitor works, it indicates that the PC and video adapter are working properly. Reconnect the suspect monitor to the PC and turn up the brightness (and contrast if necessary). You should see a faint white haze covering the display. This is the raster generated by the normal sweep of an electron beam. Remember that the PC *must* be on and running. Without the horizontal and vertical retrace signals provided by the video adapter, there will be no raster.

For a color image to fail completely, all three video drive circuits will have to be disabled. You should check all connectors between the video PC board and the main PC board. A loose or severed wire can interrupt the voltage(s) powering the board. You should also check each output from your power supply. A low or missing voltage can disable your video circuits as effectively as a loose connector. If you find a faulty supply output, you can attempt to troubleshoot the supply as discussed in Chapter 6 or you can replace the power supply outright. For monitors that incorporate the power supply onto the main PC board, the entire raster board will have to be replaced.

If supply voltage levels and connections are intact, use an oscilloscope to trace the video signals through their respective amplifier circuits. Chances are that you will see all three video signals fail at the same location of each circuit. This is usually due to a problem in common parts of the video circuits. In the example video drive board of Fig. 8-3, such common circuitry involves the components marked with 8xx numbers (e.g., Q801). If you do not have the tools or inclination to perform such component-level troubleshooting, replace the video amplifier PC board.

You should also suspect a problem with the raster blanking circuits. During horizontal and vertical retrace periods, video signals are cut off. If visible raster lines appear in your image, check the blanking signals. If you are unable to check the blanking signals, try replacing the video drive PC board. If a new video drive board fails to correct the problem, replace the main PC board.

If you should find that all three video signals check correctly all the way to the CRT (or replacing the video drive circuit does not restore the image), you should suspect a major fault in the CRT itself because there is little else that can fail. If you have a CRT tester-rejuvenator available as shown in Chapter 3, you should test the CRT thoroughly for shorted grids or a weak cathode. If the problem cannot be rectified through rejuvenation (or you do not have access to a CRT tester), try replacing the CRT. Keep in mind, however, that a CRT is usually the most expensive part of the color monitor. If each step up to now has not restored your image, you should weigh the economics of replacing the CRT versus scrapping it in favor of a new or rebuilt unit.

Symptom 5 *A single horizontal line appears in the middle of the display.* The horizontal sweep is working properly, but there is no vertical deflection. A fault has almost certainly developed in the vertical drive circuit (refer to Fig. 8-4). Use your oscilloscope to check the sawtooth wave being generated by the vertical oscillator/amplifier IC (pin 6 of IC402). A typical sawtooth wave is illustrated in Fig. 8-7. If the sawtooth wave is missing, the fault is almost certainly in the IC. For the circuit of Fig. 8-4, try replacing

IC402. If the sawtooth wave is available on IC402 pin 6, you should suspect a defect in the horizontal deflection yoke itself or one of its related components. If you are not able to check signals to the component level, simply replace the monitor's main PC board.

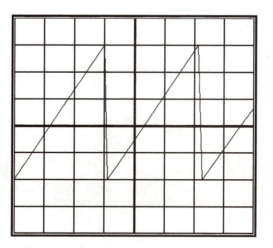

■ **8-7** *An example of the vertical sawtooth signal.*

Symptom 6 *Only the upper or lower half of an image appears.* In most cases, there is a problem in the vertical amplifier. For the example circuit of Fig. 8-4, the trouble is likely located in the vertical oscillator/amplifier (IC402). Use your oscilloscope to check the sawtooth waveform leaving IC402 pin 6. If the sawtooth is distorted, replace IC402. If the sawtooth signal reads properly, check for other faulty components in the vertical deflection yoke circuit. If you do not have the tools or inclination to check and replace devices at the component level, replace the monitor's main PC board. When the image is restored, be sure to check vertical linearity as described in Chapter 5.

Symptom 7 *A single vertical line appears along the middle of the display.* The vertical sweep is working properly, but there is no horizontal deflection. However, in order to even see the display at normal brightness, there must be high voltage present in the monitor—the horizontal drive circuit must be working (refer to Fig. 8-4). The fault probably lies in the horizontal deflection yoke. Check the yoke and all wiring connected to it. It may be necessary to replace the horizontal deflection yoke or the entire yoke assembly. Keep in mind that yoke replacement will require you to perform some careful alignment, especially dynamic convergence.

If horizontal deflection is lost as well as substantial screen brightness, there may be a marginal fault in the horizontal drive circuit. If there is a problem with the horizontal oscillator pulses, the switching characteristics of the horizontal amplifier will change. In turn, this affects high-voltage development and horizontal deflection. Use your oscilloscope to check the square wave generated by the horizontal oscillator IC301 pin 3 as shown in Fig. 8-4. You should see a wave similar to the one shown in Fig. 8-8. If the wave is distorted, replace the oscillator IC (IC301). If the horizontal pulse is correct, check the horizontal switching transistors (Q301 and Q302). Replace any transistor that appears defective. If the collector signal at the HOT is low or distorted, there may be a short circuit in the flyback transformer primary winding. Try replacing the FBT. If you do not have the tools or inclination to check components to the component level (or the problem persists), replace the monitor's main PC board. When the repair is complete, check the horizontal linearity and size as described in Chapter 5.

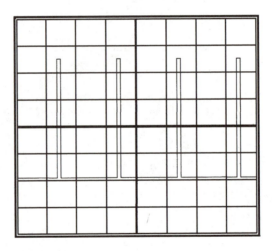

■ **8-8** *An example of the horizontal square wave pulse.*

Symptom 8 *There is no image and no raster.* When the monitor is properly connected to a PC, a series of text information should appear as the PC initializes. We can use this as our baseline image. Isolate the monitor by trying an efficiently running monitor on your host PC. If this good monitor works, it indicates that the PC and video adapter are working properly. Reconnect the suspect monitor to the PC and turn up the brightness (and contrast if necessary). Start by checking for the presence of horizontal and vertical synchronization pulses. If pulses are absent, no raster will be generated. If sync pulses are present, there is likely a problem somewhere in the horizontal drive or high-voltage circuits.

Always suspect a power supply problem and check every output from the supply (especially the 20 and 135 V dc outputs as shown in Fig. 8-4). A low or absent supply voltage will disable the horizontal deflection and high-voltage circuits. If one or more supply outputs are low or absent, you can troubleshoot the power supply circuit as discussed in Chapter 6 or replace the power supply outright (when the power circuit is combined on the monitor's main PC board, the entire main PC board will have to be replaced).

If the supply outputs read correctly, suspect your horizontal drive circuit. Use your oscilloscope to check the horizontal oscillator output at the base of Q301 as shown in Fig. 8-4. You should see a signal similar to the one in Fig. 8-8. If the square wave is low, distorted, or absent, replace the horizontal oscillator IC (IC301). If a regular pulse is present, the horizontal oscillator is working. Since Q301 is intended to act as a switch, you should also find a pulse at the collector of Q301. If the pulse output is severely distorted or absent, Q301 is probably damaged (remove Q301 and test it). If Q301 reads as faulty, it should be replaced. If Q301 reads good, check the horizontal coupling transformer (T303) for shorted or open windings. Try replacing T303 (there is little else that can go wrong in this part of the circuit).

Check the HOT (Q302) next by removing it from the circuit and testing it. If Q302 reads faulty, it should be replaced with an exact replacement part. If Q302 reads good, the fault probably lies in the flyback transformer. Try replacing the FBT. If you do not have the tools or inclination to perform these component-level checks, simply replace the monitor's main PC board outright.

In the event that these steps fail to restore the image, the CRT has probably failed. If you have access to a CRT tester-rejuvenator as discussed in Chapter 3, you can test the CRT. When the CRT measures as bad (and cannot be restored through rejuvenation), it should be replaced. If you do not have a CRT test instrument, you can simply replace the CRT. Keep in mind, however, that a CRT is usually the most expensive part of a color monitor. If each step up to now has not restored your image, you should weigh the economics of replacing the CRT versus scrapping it in favor of a new or rebuilt unit. If you choose to replace the CRT, you should perform a full set of alignments as described in Chapter 5.

205

Symptom 9 *The image is too compressed or too expanded. A whitish haze may appear along the bottom of the image.* Start by checking your vertical size control to be sure that it was not adjusted accidentally. Since vertical size is a function of the vertical sawtooth oscillator, you should suspect the vertical oscillator circuit. A sawtooth signal that is too large will result in an over-expanded image, whereas a signal that is too small will appear to compress the image. Use your oscilloscope to check the sawtooth signal. For the vertical drive circuit of Fig. 8-4, the signal on IC402 pin 6 should appear similar to the one shown in Fig. 8-7. If the signal is incorrect, try replacing IC402. You may also wish to check the PC board for any cracks or faulty soldering connections around the vertical oscillator circuit. If the problem persists, or you do not have the tools or inclination to perform component-level troubleshooting, simply replace the monitor's main PC board outright.

Symptom 10 *The displayed characters appear to be distorted.* The term *distortion* can be interpreted in many different ways. For our purposes, we will simply say that a distorted image (usually text) is difficult to read. Before even opening your toolbox, check the monitor's location. The presence of stray magnetic fields in close proximity to the monitor can cause bizarre forms of distortion. Try moving the monitor to another location. Remove any electromagnetic or magnetic objects (e.g., motors or refrigerator magnets) from the area. If the problem persists, the monitor is likely at fault.

If only certain areas of the display appear affected (or affected worse than other areas), the trouble is probably due to poor linearity (either horizontal, vertical, or both). If raster speed varies across the display, the pixels in some areas of the image may appear too close together, while the pixels in other areas of the image may appear too far apart. You can check and correct horizontal and vertical linearity using the procedures found in Chapter 5. Alignment test patterns are available on the companion disk for this book. If alignment fails to correct poor linearity, your best course is often simply to replace the monitor's main PC board.

If the image is difficult to read because it is out of focus, you should check the focus alignment using the procedure outlined in Chapter 5. If you cannot achieve a sharp focus using controls either on the front panel of the monitor or on the flyback transformer assembly, there is probably a fault in the flyback transformer. Try replacing the FBT. If the problem persists, your best course is often simply to replace the monitor's main PC board.

Symptom 11 *The display appears wavy.* There are visible waves appearing along the edges of the display as the image sways back and forth. This is almost always the result of a power supply problem—one or more outputs are failing. Use your multimeter and check each supply output. If you find a low or absent output, you can proceed to troubleshoot the supply as discussed in Chapter 6 or you can simply replace the supply outright. If the power supply is integrated onto the main PC board, you will have to replace the entire main PC board.

Symptom 12 *The display is too bright or too dim.* Before opening the monitor, be sure to check the brightness and contrast controls. If the controls had been accidentally adjusted, set contrast to maximum and adjust the brightness level until a clear, crisp display is produced. When front panel controls fail to provide the proper display (but focus seems steady), suspect a fault in the monitor's power supply. Refer to the example schematic of Fig. 8-4. If the 135-V dc supply is too low or too high, brightness levels controlling the CRT screen grid will shift. If you find one or more incorrect outputs from the power supply, you can troubleshoot the supply as discussed in Chapter 6 or replace the supply outright. For those monitors that incorporate the power supply on the main PC board, the entire main PC board will have to be replaced. If both brightness and focus seem out, there is still probably a voltage problem, and you should refer to Symptom 10.

Symptom 13 *You see visible raster scan lines in the display.* The very first thing you should do is check the front-panel brightness and contrast controls. If contrast is set too low and/or brightness is set too high, raster will be visible on top of the image. This will tend to make the image appear a bit fuzzy. If the front-panel controls cannot eliminate visible raster from the image, chances are that you have a problem with the power supply. Use your multimeter and check each output from the supply. If one or more outputs appear too high (or too low), you can troubleshoot the supply as described in Chapter 6 or replace the supply outright. If the supply is integrated with the monitor's main PC board, the entire PC board will have to be replaced.

If the power supply is intact, you should suspect a problem with the raster blanking circuits. During horizontal and vertical retrace periods, video signals are cut off. If visible raster lines appear in your image, check the blanking signals. If you are unable to check the blanking signals, try replacing the video drive PC board. If a new video drive board fails to correct the problem, replace the main PC board.

Symptom 14 *Colors bleed or smear.* Ultimately, this symptom occurs when unwanted pixels are excited in the CRT. However, it can be caused by several different problems. Perhaps the most common problem is a fault in the video cable between the video board and the monitor. Electrical noise (sometimes called *crosstalk*) in the cable may allow signals representing one color to accidentally be picked up in another color signal wire. This can easily cause unwanted colors to appear in the display. Although the video cable is designed to be shielded and carefully filtered, age or poor installation can precipitate this type of problem. Try wiggling the cable. If the problem stops, appears intermittent, or shifts around, you have likely found the source of the problem and should replace the cable with a proper replacement assembly.

If the video cable appears intact, suspect failing capacitors in the video amplifier circuits. You can see these capacitors in the schematic of Fig. 8-6. Capacitors such as C107 and C109 are typically low-value, high-voltage components (i.e., C107 is 10 µF at 160 V, C108 is 1 µF at 150 V, and C109 is 10 µF at 50 V), so they tend to degrade rather quickly. Fortunately, such capacitors are easy to spot on the video amplifier board. If the color problem appears intermittent (or occurs when the monitor warms up), try a bit of liquid refrigerant on each capacitor. If the problem disappears, the one you froze is probably defective. Otherwise, you can turn off and unplug the monitor; then check each capacitor individually. When replacing capacitors in the video amplifier circuit, be sure to replace them with the same type and voltage rating.

If capacitors are not at fault, suspect the amplifier transistors on the video amplifier board (i.e., Q101, Q102, Q103, or Q104). Turn off and unplug the monitor; then try checking each of the transistors. Chances are that your readings will be inconclusive, so try comparing readings from each transistor to find a device that gives the most unusual readings. Try replacing any defective or questionable amplifier transistors. If you do not have the time or inclination to troubleshoot the video amplifier board, try replacing the board outright.

Symptom 15 *Colors appear to change when the monitor is warm.* Either colors will appear correctly when the monitor is cold and then change as the monitor warms up or vice versa. In both cases, there is likely to be some kind of thermal problem in the video amplifier circuits. Turn off and unplug the monitor; then start by checking the video cable, especially its connection to the raster board inside the monitor. If this connection is loose, it may be intermittent or unreliable. Tighten any loose connections and try the monitor again. Also check the cable that connects the video amplifier board to the raster board.

If the connections appear tight, your best course of action is often to remove the video amplifier board and try resoldering each of the junctions. Chances are that age or thermal stress has fatigued one or more connections. By resoldering the connections, you should be able to correct any potential connection problems. You might also try resoldering the connector which passes video data from the raster board to the video amplifier board. If your problems persist, try replacing the video amplifier board.

Symptom 16 *An image appears distorted in 350 or 400 line mode.* In most cases, the distortion is an image that appears excessively compressed. As you probably read earlier in this book, different screen modes have a different number of horizontal lines (i.e., a 640×480 display offers 480 horizontal traces of 640 pixels each). When the screen mode changes, the number of lines changes as well (i.e., to a 320×200 mode). As you might expect, the "size" of each pixel has to be adjusted when the screen mode changes in order to keep the image roughly square; otherwise, the image simply "shrinks." Monitors detect the screen mode by checking the polarity of the sync signals. You can see this function in the schematic of Fig. 8-6 (in the block representation of IC201).

Typically, each screen mode size can be optimized by an adjustment on the raster board. However, if a mode adjustment is thrown off (or the sync sensing circuit fails), an image can easily appear with an incorrect size. If you notice this kind of distortion without warning, suspect a problem with the sync sensing circuit. If the sync sensing circuit is incorporated into a single IC (e.g., IC201), replace the IC outright. If you notice a size problem after aligning the monitor, you may have accidentally upset a size adjustment. Readjust the size controls to restore proper image dimensions.

Symptom 17 *The fine detail of high-resolution graphic images appears a bit fuzzy.* At best, this kind of symptom may not appear noticeable without careful inspection, but it may signal a serious problem in the video amplifier circuit. High resolutions demand high bandwidth—a video amplifier must respond quickly to the rapid variations between pixels. If a weakness in the video amplifier(s) occurs, it can limit bandwidth and degrade video performance at high resolutions. The problem will probably disappear at lower resolutions.

The particular problem with this symptom is that it is almost impossible to isolate a defective component—the video amplifier board *is* working. As a result, your best course of action is to check all connectors for secure installation first. Nicked or frayed video cables can also contribute to the problem. If the problem remains, replace the video amplifier board.

Symptom 18 *The display changes color, flickers, or cuts out when the video cable is moved.* Check the video cable's connection to the video adapter at the PC. A loose connection will almost certainly result in such intermittent problems. If the connection is secure, there is an intermittent connection in the video cable. Before replacing the cable, check its connections within the monitor itself. When connections are intact, replace the intermittent video cable outright. Do not bother cutting or splicing the cable—any breaks in the signal shielding will cause crosstalk which will result in color bleeding.

Symptom 19 *The image expands in the horizontal direction when the monitor gets warm.* One or more components in the horizontal retrace circuit are weak and are slightly changing value once the monitor gets warm. Turn off and unplug the monitor. You should inspect any capacitors located around the horizontal output transistor (HOT). Typical suspect components would include C410 and C411 (from Fig. 8-6). The problem is that thermal problems such as this can be extremely difficult to isolate because you can't measure capacitor values while the monitor is running, and after the monitor is turned off, the parts will cool too quickly to catch a thermal problem. It is often most effective simply to replace several of the key capacitors around the HOT outright. If you don't want to bother with individual components, replace the raster board.

Symptom 20 *The image shrinks in the horizontal direction when the monitor gets warm.* This is another thermal-related problem which indicates either a weakness in one or more components or a mild soldering-related problem. Turn off and unplug the monitor. Start by checking for a poor solder connection, especially around the horizontal deflection yoke wiring, the horizontal output transistor (HOT), and the FBT. If nothing appears obvious, you might consider resoldering all of the components in the HOT area of the raster board.

If problems continue, suspect a failure in the HOT itself. Semiconductors rarely become marginal; they either work or they don't. Still, semiconductor junctions can become unstable when temperatures change and result in circuit characteristic changes. You could also try replacing the HOT outright.

It is also possible that one or more midrange power supply outputs (i.e., 12 or 20 V) are sagging when the monitor warms up. Use a voltmeter and measure the outputs from your power supply. If the 12- or 20-V outputs appear to drop slightly once the monitor has been running for a bit, you should troubleshoot the power supply as described in Chapter 6.

Symptom 21 *High voltage fails after the monitor is warm.* There are a large number of possible causes behind this problem, but no matter what permutation you find, you will likely be dealing with soldering problems or thermal-related failures. Turn off and unplug the monitor. Inspect the HOT's heat sink assembly. There may be a bad solder connection on the heat sink ground. There may also be an open solder connection on one or more of the FBT pins. If you cannot locate a faulty soldering connection, you may simply choose to resolder all of the connections in the flyback area.

If the problem persists, you should suspect that either your HOT or FBT is failing under load (after the monitor warms up). One possible means of isolating the problem is to measure pulses from the HOT output with your oscilloscope. If the pulses stop at the same time your high voltage fails, you can suspect a problem with your HOT or other horizontal components. Try replacing the HOT. If high voltage fails but the HOT pulses remain, your FBT has probably failed. Replace the FBT. If you do not have an oscilloscope, try replacing the HOT first because that is the least expensive part; then replace the FBT if necessary.

In the unlikely event that both a new HOT and FBT do not correct the problem, you should carefully inspect the capacitors in the HOT circuit. One or more might be failing. Unfortunately, it is very difficult to identify a marginal capacitor (especially one that is suffering from a thermal failure). You may try replacing the major capacitors in the HOT circuit or replace the raster board entirely.

Symptom 22 *The image blooms intermittently.* The amount of high voltage driving the CRT is varying intermittently. Since high voltage is related to the HOT circuit and FBT, you should concentrate your search in those two areas of the raster board. Examine the soldering of your HOT and FBT connections, especially the ground connections if you can identify them. You may try resoldering all of the connections in those areas (remember to turn off and unplug the monitor before soldering). There may also be a ground problem on the video amplifier board which allows all three color signals to vary in amplitude. When this happens, the overall brightness of the image changes, and the image may grow or shrink a bit in response. Try resoldering connections on the video amplifier board.

If the problem remains (even after soldering), your FBT may be failing, probably due to an age-related internal short. High-end test equipment such as Sencore's monitor test station provides the instrumentation to test an FBT. If you do not have access to such dedicated test equipment, however, try replacing the FBT assembly. If you do not have the time or inclination to deal with component replacement, go ahead and replace the raster board outright. In the unlikely event that your problem persists, suspect a fault in the CRT itself. If you have access to a CRT tester-rejuvenator, you can check the CRT's operation. Some weaknesses in the CRT may be corrected (at least temporarily) by rejuvenation. If the fault cannot be corrected, you may have to replace the CRT.

Symptom 23 *The image appears out of focus.* Before suspecting a component failure, try adjusting the focus control. In most cases, the focus control is located adjacent to the FBT. Keep in mind that the focus control should be adjusted with brightness and contrast set to optimum values—excessively bright images may lose focus naturally. If the focus control is unable to restore a proper image, check the CRT focus voltage. In Fig. 8-8, you can find the focus voltage off an FBT tap. If the focus voltage is low (often combined with a dim image), you may have a failing FBT. It is possible to test the FBT if you have the specialized test instrumentation; otherwise, you should just replace the FBT outright. If you lack the time or inclination to replace the FBT, you can simply replace the raster board.

If a new FBT does not resolve your focus problem, suspect a fault in the CRT, probably in the focus grid. You can use a CRT tester-rejuvenator to examine the CRT, and it may be possible to restore normal operation (at least temporarily). If you do not have such equipment, you will simply have to try a new CRT.

Symptom 24 *The image appears to flip or scroll horizontally.* There is a synchronization problem in your horizontal raster circuit. Begin by checking the video cable to be sure that it is installed and connected securely. Cables that behave intermittently (or that appear frayed or nicked) should be replaced. If the cable is intact, suspect a problem in your horizontal circuit. If there is a horizontal sync (or "horizontal hold") adjustment on the raster board, adjust it in small increments until the image snaps back into sync. If there is no such adjustment on your particular monitor, try resoldering all of the connections in the horizontal processing circuit. If the problem persists, replace the horizontal oscillator IC or replace the entire raster board.

Symptom 25 *The image appears to flip or scroll vertically.* There is a synchronization problem in your vertical raster circuit. Begin by checking the video cable to be sure that it is installed and connected securely. Cables that behave intermittently (or that appear frayed or nicked) should be replaced. If the cable is intact, suspect a problem in your vertical circuit. If there is a vertical sync (or "vertical hold") adjustment on the raster board, adjust it in small increments until the image snaps back into sync. If there is no such adjustment on your particular monitor, try resoldering all of the connections in the vertical processing circuit. If the problem persists, replace the vertical oscillator IC or replace the entire raster board.

Symptom 26 *The image appears to shake or oscillate in size.* This may occur in bursts, but it typically occurs constantly. In most cases, it is due to a fault in the power supply, usually the 135-V (B+) output. Try measuring your power supply outputs with an oscilloscope and see if an output is varying along with the screen size changes. If you locate such an output, the filtering portion of that output may be malfunctioning. Track the output back into the supply and replace any defective components. If you are unable to isolate a faulty component, replace the power supply. When the power supply is integrated onto the raster board, you may have to replace the raster board entirely.

213

If the outputs from your power supply appear stable, you should suspect a weak capacitor in your horizontal circuit. Try resoldering the FBT, HOT, and other horizontal circuit components to eliminate the possibility of a soldering problem. If the problem remains, you will have to systematically replace the capacitors in the horizontal circuit. If you do not have the time or inclination to replace individual components, replace the raster board outright.

Here's an unusual problem. The shaking you see may be related to a problem in the degaussing coil located around the CRT funnel. Ordinarily, the degaussing coil should unleash most of its energy in the initial moments after monitor power is turned on. Thermistors (or posistors) in the power supply quickly diminish coil voltage, effectively cutting off the degaussing coil's operation. A fault in the degaussing coil circuit (in the power supply) may continue to allow enough power to the coil to affect the image's stability. Try disconnecting the degaussing coil. If the problem remains, the degaussing coil is likely operating properly. If the problem disappears, you have a fault in the degaussing coil circuit.

9

Flat-panel displays

TRADITIONAL COMPUTER MONITORS HAVE BEEN BASED ON the CRT. While CRTs have evolved to meet the demands of improved resolution and color depth, they are still limited by their unwieldy size, weight, and heavy power consumption, which are certainly undesirable factors for mobile computing. *Flat-panel displays* offer the small size, light weight, reasonably low power consumption, and general ruggedness that are ideal for notebook, palmtop, and pen systems (Fig. 9-1). Concerns about the environment and monitor emissions have even pushed flat-panel displays into use as desktop monitors. No book on computer monitors would be complete without a chapter devoted to flat-panel displays. Two families of flat-panel displays have evolved for use in small computers: liquid crystal displays (LCDs) and gas plasma displays (GPDs). This chapter will show you the technology behind LCDs and GPDs and present a series of troubleshooting procedures.

Flat-panel display characteristics

Before jumping right into detailed discussions of flat-panel display technologies, it would be helpful for you to have a clear understanding of a display's major characteristics. This part of the chapter also presents a set of cautions for display handling. Even if you are already familiar with flat-panel specifications, take a moment to review their handling precautions.

Pixel organization

As with CRT-based displays, the images formed on a flat-panel display are *not* solid images. Instead, images are formed as an array of individual picture elements (*pixels*). Pixels are arranged into a matrix of rows (top to bottom) and columns (left to right) as illustrated in Fig. 9-2. Each pixel corresponds to a location in video RAM (not core memory, which holds programs and data). As data are written into video RAM, pixels in the array will turn on and off.

■ **9-1** *A Sharp PC8650 notebook computer. (Sharp Electronics)*

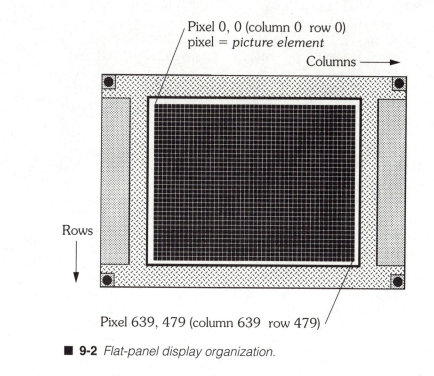

Pixel 0, 0 (column 0 row 0)
pixel = *picture element*

Columns ———————▶

Rows

Pixel 639, 479 (column 639 row 479)

■ **9-2** *Flat-panel display organization.*

The on/off patterns that appear in the array form letters and graphics.

The resolution of a flat-panel display is little more than the number of pixels that can be displayed. More pixels allow the display to present finer, higher quality images. As an example, many notebook computer displays are capable of showing 307,200 dots arranged in a standard VGA array of 640 columns by 480 rows ($640 \times 480 = 307,200$), whereas older laptop and notebook displays are capable of only 128,000 pixels arranged as 640 columns by 200 rows (or less). The very newest systems employ displays capable of handling 480,000 pixels in an 800×600 format. As time goes on, flat-panel displays will approach the high resolutions available in current monitors.

Aspect ratio

The *aspect ratio* is basically the "squareness" of each pixel and, indirectly, the squareness of the display. For example, a display with perfectly square pixels has an aspect ratio of 1:1. A rectangle box 100 pixels wide and 100 pixels high would appear as an even square. However, pixel aspect ratios are not always 1:1. Typical pixels are somewhat higher than they are wide. For a display with 320×200 resolution, a pixel width of 0.34 mm and height of 0.48 mm (1:1.41) is not uncommon. Higher resolution displays use smaller dots to fit more pixels into roughly the same viewing area. As a result, smaller pixels tend to approach 1:1 aspect ratios. Keep in mind that aspect ratio may not be shown as an individual display specification. Figure 9-3 illustrates the concept of aspect ratio.

Viewing angle

Every display has a particular *viewing angle*. It is the angle through which a display can be viewed "clearly" as shown in Fig. 9-4. Viewing angle is rarely a concern for bright, crisp displays such as CRTs and gas plasma flat panels. Such displays generate light, so they can be seen up to a very wide angle (usually up to 70° from center). For LCDs, viewing angle is a critical specification. Liquid crystal displays do not generate their own light, so the display contrast tends to degrade quickly as you leave a direct line of sight.

Position yourself or your small computer so that you look directly at the display (a perpendicular orientation). Tilt the screen up and away from you. Do you see how the contrast and brightness of the display decrease as you tilt the screen? The angle of the display when the picture just becomes indiscernible is the negative verti-

216

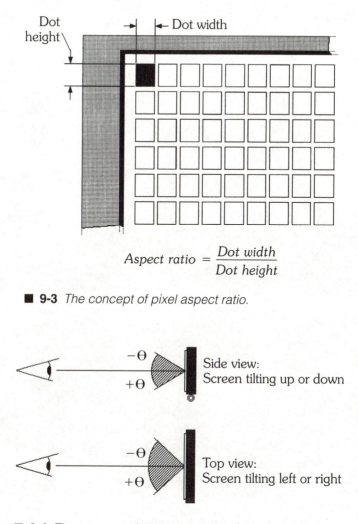

Dot height

Dot width

$$\text{Aspect ratio} = \frac{\text{Dot width}}{\text{Dot height}}$$

■ **9-3** *The concept of pixel aspect ratio.*

$-\Theta$
$+\Theta$
Side view:
Screen tilting up or down

$-\Theta$
$+\Theta$
Top view:
Screen tilting left or right

■ **9-4** *The concept of display viewing angle.*

cal limit ($-\Theta$). Return the screen to a direct line of sight and then slowly tilt the screen down toward you. Once again, you will see the display degrade. The angle of the display when the display becomes indiscernible is the positive vertical limit ($+\Theta$). Ideally, both vertical limits should be the same. Return the display to a direct line of sight; then repeat this test in the horizontal orientation. Swing the display left or right until the image becomes indiscernible. These are the negative and positive horizontal limits ($-\phi$ and $+\phi$), respectively. Ideally, both horizontal limits should be the same. As a general rule, the larger a viewing angle is, the easier the display is to look at.

Contrast

The *contrast* of an image is loosely defined as the difference in luminous intensity between pixels that are fully on and pixels that are fully off. The greater this difference is, the higher the contrast is, and the image appears sharper. Many graphic flat-panel LCDs offer contrast ranging from a low of 4 to 9 and higher. For the purposes of this book, contrast is a unitless number. Since luminous intensity is strongly dependent on a display's particular viewing angle, a reference angle is typically added to the contrast specification—usually a straight-on view.

Remember that contrast is a comparison of black versus white. It is desirable (especially with monochrome displays) to simulate 16, 32, 64, or more gray levels that are somewhere *between* black and white. Any gray level other than pure black provides a lower contrast versus white. Do not confuse gray scale levels with poor contrast.

Response time

The *response time* of a display is the time required for a display pixel to reach its on or off condition after the pixel has been addressed by the corresponding driver circuitry. Such on/off transitions do not occur instantaneously. Depending on the vintage and quality of the display, pixel response times can vary anywhere from 40 to 200 ms. Gas plasma and active matrix liquid crystal displays typically offer the shortest response times, while older passive matrix LCDs provide the slowest performance. You will see much more about the various display types and techniques a bit later in this chapter.

Handling precautions

Next to magnetic hard drives, flat-panel displays are some of the most sophisticated and delicate assemblies in the computer industry. While most gas plasma and liquid crystal displays can easily withstand the rigors of everyday use, there are serious physical, environmental, and handling precautions that you should be aware of before attempting any sort of repair.

You must be extremely careful with **all** liquid crystal (LC) assemblies. Liquid crystal material is sandwiched between two layers of fragile glass. The glass can easily be fractured by abuse or careless handling. If a fracture should occur and liquid crystal material happens to leak out, use rubber gloves and wipe up the spill with soap and water. Immediately wash off any LC material that comes in contact with your skin. *Do not, under any circumstances, ingest or inhale LC material.* Avoid applying pressure to the surface of an LCD. You risk scratching the delicate polarizer layer covering the display's face. If the polarizer is scratched or damaged in any way, it will have to be replaced. Excessive pressure or bending forces can fracture the delicate connections within the display.

You can use *very* gentle pressure to clean the face of a display. Lightly wet a soft (lintfree), clean cloth with fresh isopropyl alcohol or ethyl alcohol; then **gently** wipe away the stain(s). You may prefer to use photography-grade lens wipes instead of a cloth. *Never* use water or harsh solvents to clean a display. Water drops can accumulate as condensation, and high-humidity environments can corrode a display's electrodes. When a display assembly is removed from your system, keep the assembly in an antistatic bag in a dry, room-temperature environment. *Never* store your small computer in a cold vehicle, room, or other environment. Liquid crystal material coagulates (becomes firm) at low temperatures (below 0° C or 32° F). Exposure to low temperatures can cause black or white bubbles to form in the material.

Flat-panel display circuitry is very susceptible to damage from electrostatic discharge (ESD). Make certain to use antistatic bags or boxes to hold the display. Use an antistatic wrist strap to remove any charges from your body. Use a grounded soldering iron and tools. Work on an antistatic workbench mat if possible. Try to avoid working in cool, dry environments where static charges can accumulate easily. If you use a vacuum cleaner around your display, make sure that the vacuum is static safe.

Liquid crystal display technology

Liquid crystal (LC) is an unusual organic material that has been known by science for many years. Although it is liquid in form and appearance, LC exhibits a crystalline molecular structure that resembles a solid. If you were to look at a sample of LC material under a microscope, you would see a vast array of rod-shaped molecules. In its normal state, LC is virtually clear; light would pass right through a container of LC. When LC material is assembled into a flat panel, the molecules have a tendency to twist.

Quite by accident, it was discovered that a voltage applied across a volume of LC forces the molecules between the active electrodes to straighten. When the voltage is removed, the straightened LC molecules return to their normal twisted orientation. It would have been a simple matter to have dismissed the LC effect as little more than a scientific curiosity, but further experiments revealed an interesting phenomenon when light polarizing materials (or *polarizers*) are placed on both sides of the LC layer; areas of the LC material that are excited by an external voltage become dark and visible. When voltage is removed, the area becomes clear and invisible again. A *polarizer* is a thin film which allows light to pass in only one orientation.

By using electrodes with different patterns, various images can be formed. The earliest developments in commercial LCDs were simply referred to as *twisted nematic* (TN) displays. A typical LCD assembly is illustrated in Fig. 9-5. Notice that an array of transparent electrodes are printed and sealed on the inside of each glass layer.

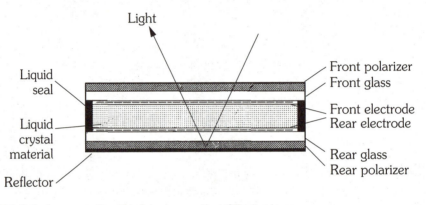

Light

Liquid seal

Liquid crystal material

Reflector

Front polarizer
Front glass

Front electrode
Rear electrode

Rear glass
Rear polarizer

■ **9-5** *A conventional twisted nematic LCD display.*

There are four major varieties of liquid crystal assemblies that you should be familiar with: twisted nematic (TN), super twisted nematic (STN), neutralized super twisted nematic (NTN or NSTN), and film-compensated super twisted nematic (FTN or FSTN). Each of these variations handles light somewhat differently and offers unique display characteristics.

TN LCDs

The *twisted nematic* (TN) display is illustrated in Fig. 9-6. Light can originate from many different sources and strike the front polarizer, but the vertically oriented polarizer only allows light waves in the vertical orientation to pass through into the LC cell. As vertically oriented light waves enter the LC assembly, their orientation twists 90° following the molecular twist in the LC material. As light leaves the LC cell, its orientation is now horizontal. Since the rear polarizer is aligned horizontally, light passes through and the LC display appears transparent.

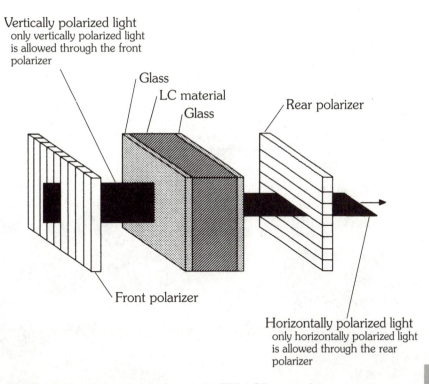

Vertically polarized light
only vertically polarized light
is allowed through the front
polarizer

Glass

LC material

Glass

Rear polarizer

Front polarizer

Horizontally polarized light
only horizontally polarized light
is allowed through the rear
polarizer

■ **9-6** *Light path in a twisted nematic (TN) LCD assembly.*

When a pixel is activated, the LC material being energized straightens its alignment. Twist will become 0°, and light will not change its polarization in the LC cell. Vertically polarized light is blocked by the horizontally oriented rear polarizer. This makes the activated pixel appear dark.

Twisted nematic technology is appealing for its low cost, simple construction, and good response time, but it is limited by poor viewing angle and low contrast in high-resolution displays. Today, TN displays have been largely replaced by any one of the three following technologies.

STN LCDs

A *super twisted nematic* (STN) assembly is shown in Fig. 9-7. Initially, the STN approach appears identical to the TN technique, but there are two major differences. First, a "super" twisted LC material is used which provides more than 200° of twist instead of only 90° in the TN formulation. The rear polarizer angle must also be changed to match the twist of the LC material. For example, if the LC material has a twist of 220°, the rear polarizer must be aligned to that same orientation.

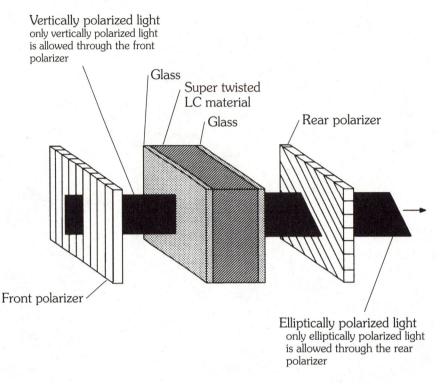

Vertically polarized light
only vertically polarized light
is allowed through the front
polarizer

Glass

Super twisted
LC material

Glass

Rear polarizer

Front polarizer

Elliptically polarized light
only elliptically polarized light
is allowed through the rear
polarizer

■ **9-7** *Light path in a super twisted nematic (STN) LCD assembly.*

In STN operation, the vertically oriented light passing through the front polarizer enters the LC cell. As light passes through the LC cell, its orientation changes following the formulation's particular twist. The twist may be as little as 200° or as much as 270°. Light leaving the LC cell should then pass through the customized rear polarizer and make the display appear transparent. If a pixel is activated, the LC material at that point will straighten completely. Light no longer twists to match the rear polarizer, so the pixel appears dark.

An STN display offers much better contrast and viewing angle than TN versions because of the additional twist. Super twisted nematic technology also performs very well at high resolutions (up to 1024 × 800 pixels). However, STN displays cost more than regular TN displays, and the response time to activate each pixel is somewhat slow because of the extra twist.

NTN LCDs

A *neutralized super twisted nematic* (NTN or NSTN) display is shown in Fig. 9-8. Light is vertically oriented by the front polarizer before being admitted to the first LC cell. Light entering the first LC cell is twisted more than 270°. A second LC cell (known as a *compensator cell*) adds extra twist to light polarization, resulting in a horizontally oriented light output. Light that passes through the second LC cell also passes through the rear polarizer and results in a clear (transparent) display. Keep in mind that only the first LC cell offers active pixels. The compensator cell only adds twist.

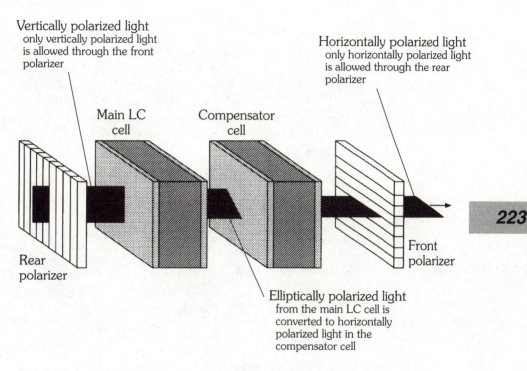

9-8 *Light path in a neutralized STN (NSTN or NTN) LCD assembly.*

When a pixel is activated in the first LC cell, the LC molecules align, and light at that point is not twisted. The untwisted light is not twisted enough by the compensator cell, so that point is blocked by the rear polarizer and appears dark. Light passing through an idle pixel is twisted and then twisted again by the compensator cell. With this additional twist, light passes through the rear polarizer and the deactivated points appear transparent.

Neutralized super twisted nematic displays produce some of the finest, high-contrast, high viewing angle images available, but

NTN displays are also much heavier, thicker, and costlier than other types of displays. It is also difficult to backlight such a configuration of LC cells. For most small-computer applications, FTN displays are preferred over NTN models.

FTN LCDs

Figure 9-9 illustrates the basic structure of a *film-compensated super twist nematic* (FTN or FSTN) display. As you can see, the FTN display is very similar to the NTN display shown in Fig. 9-8. However, an FTN display uses a layer of optically compensated film instead of a second LC cell to achieve horizontal light polarization. Vertically oriented light passes through the front polarizer and then is twisted more than 200° by the LC cell. When light emerges from the LC cell, it passes through a compensator film. Assuming that light is oriented properly from the LC cell, the compensator layer changes light polarization to a horizontal orientation. Light then passes through the horizontal polarizer causing a clear (transparent) display.

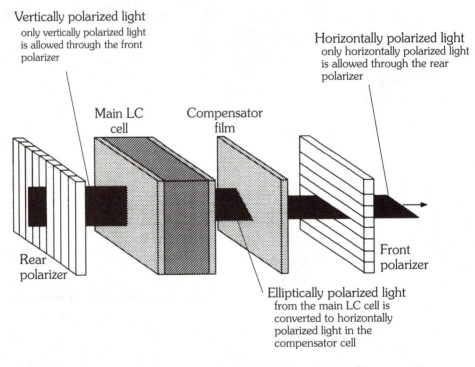

Vertically polarized light
only vertically polarized light is allowed through the front polarizer

Horizontally polarized light
only horizontally polarized light is allowed through the rear polarizer

Main LC cell

Compensator film

Rear polarizer

Front polarizer

Elliptically polarized light
from the main LC cell is converted to horizontally polarized light in the compensator cell

■ **9-9** *Light path in a film-compensated STN (FSTN or FTN) LCD assembly.*

When a pixel is activated, the LC material at that point straightens, and light polarization does not occur. As unaltered light passes through the compensation film, it does not twist enough to pass through the rear polarizer, so the pixel appears dark.

An FTN LCD is much lighter, thinner, and less expensive than its NTN counterpart. The FTN display does not have nearly as much optical loss as NTN versions, so FTN displays are easy to backlight. The only major disadvantage to an FTN display is that its contrast and viewing angle are slightly reduced because of the compensating film.

Viewing modes

It is important for you to realize that light plays a critical role in the formation of liquid crystal images. The path that light takes through the LC assembly and your eyes can have a serious impact on the display's image quality as well as the display's utility in various environments. There are three classical viewing "modes" that you should understand: reflective LCD, transflective LCD, and transmissive LCD. Figure 9-10 illustrates the action of each mode.

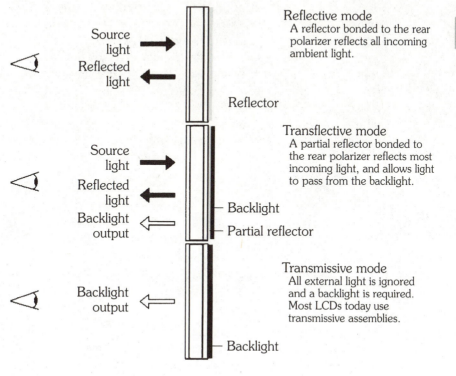

■ 9-10 *LCD viewing modes.*

In the *reflective* viewing mode, only available light is used to illuminate the display. A metallized reflector is mounted behind the display's rear polarizer. Light from the outside environment that penetrates the LC assembly is reflected back to your eyes resulting in a clear (transparent) image. Light that is blocked due to an activated pixel appears dark. Reflective displays work best when used in an outdoor or well-lit environment. For example, the LCDs used in digital multimeters are almost always operated in reflective mode. If light is blocked from the display, the image will virtually disappear. However, since no backlighting is used, the display consumes very little power.

The *transflective* viewing mode uses a partial reflector behind the LC cell's rear polarizer. This partial reflector will reflect light provided by the outside environment and pass any illumination provided from behind the assembly (the backlight). Transflective operation allows the display to be operated in direct light with the backlight turned off. The backlight can then be activated in low-light conditions.

A *transmissive* LCD uses a transparent rear polarizer with no reflector at all. Light entering the LCD assembly from the outside environment is not reflected back to your eyes. Instead, a backlight is required to make the image visible. When pixels are off, backlight illumination passes directly through the display to your eyes resulting in clear (transparent) pixels. Activated pixels block the backlight and result in dark points. The advantage to transmissive operation is that displays can be illuminated very evenly, which is important in low-light or indoor environments. Unfortunately, the backlight can be overpowered by bright light or sunlight, so the transmissive display can appear pale or "washed out" when used outdoors.

Backlighting

Backlighting is the process of adding a known light source to an LCD in order to improve the display's visibility in low-light situations. There are three primary approaches to backlighting: electroluminescent (EL) panels, cold-cathode fluorescent tubes (CCFTs), and light emitting diodes (LEDs).

Electroluminescent (EL) panels are thin, lightweight, and produce a very even light output across their surface area. These panels are available in a several colors, but white is preferred for computer displays. The EL panel is usually mounted directly behind the display's rear polarizer (or transflector if used) as shown in Fig. 9-11. These panels are reasonably rugged and reliable, but they require a substantial ac excitation voltage in order to operate. An ac inverter supply is used to convert low-voltage dc into a high-voltage ac level of 100 V ac or more. A disadvantage to EL panels is their relatively short working life (2000 to 3000 h) before a serious loss of backlight intensity occurs.

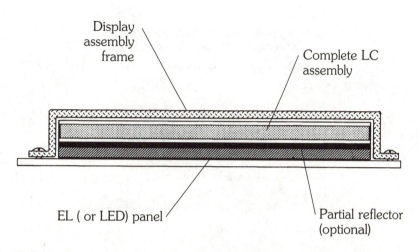

■ **9-11** *Construction of a typical EL (or LED) backlight assembly.*

An array of light-emitting diodes (LEDs) may be used for backlighting small displays. As with EL panels, LED arrays are thin and lightweight assemblies. They offer better brightness than most comparable EL panels and offer an exceptionally long working life of more than 50,000 h. Unfortunately, LED backlight assemblies draw much more power and shed more heat than EL panels, even though the LEDs will run from +5.0 V dc. Also, LEDs are not yet available in white light configurations that are favored for computer applications. Instead, most LED backlight panels produce a yellow-green color. Light-emitting diode backlights are primarily used in small character-oriented displays for such devices as fax machines or photocopiers.

Cold-cathode fluorescent tubes (CCFTs) offer a very bright source of white light that consumes reasonably little power. They also enjoy a long life (10,000 to 15,000 h) without serious degradation. Such characteristics have made CCFTs very popular in a great many notebook and pen-computer displays. As Fig. 9-12 illustrates, there are two common methods of mounting CCFTs: edge-lighting and backlighting.

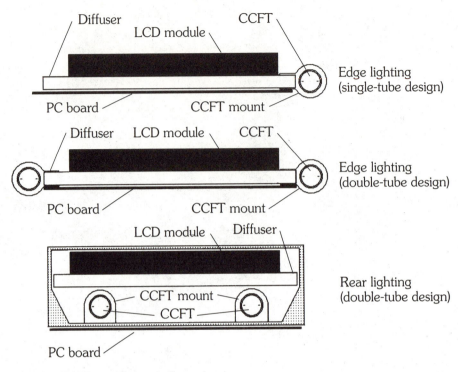

Diffuser   CCFT
LCD module

Edge lighting
(single-tube design)

PC board   CCFT mount

Diffuser   LCD module   CCFT

Edge lighting
(double-tube design)

PC board   CCFT mount

LCD module   Diffuser

Rear lighting
(double-tube design)

CCFT mount
CCFT

PC board

■ **9-12** *CCFT backlight configurations.*

As you may imagine, edgelighting is favored in thin or low-profile displays (which applies to virtually all modern small computers). A layer of translucent material referred to as the *diffuser* distributes the lamp's light evenly behind the LC cell. To create an even brighter display, a second CCFT can be added on the opposite edge of the diffuser. If a smaller, thicker display assembly is preferred, one or two CCFTs can be mounted in a cavity directly behind the LC cell. A diffuser is still used to spread light evenly behind the LC cell. Cold-cathode fluorescent tubes need a high-voltage ac source to operate, so an inverter supply is employed to provide the high voltages that most CCFTs need.

Passive and active matrix operation

Up to now, this chapter has shown you much of the background information and technology associated with monochrome (black and white) LCDs. Now that you understand what LCDs are and how they are constructed, you need to understand how each pixel in a display is controlled (or *addressed*). There are two classical methods of LCD addressing: passive addressing and active addressing. You will probably encounter both types of addressing at one time or another.

A *passive matrix* LCD is illustrated in Fig. 9-13. Each layer of glass in an LC cell contains transparent electrodes deposited on the inside of the glass sheet. The upper (or front) glass contains column electrodes, and the lower (or rear) glass is printed with row electrodes. When both sheets of glass are fitted together as shown, a matrix is formed. Every point where a row electrode and a column electrode intersect is a potential pixel. To "light" a pixel, the appropriate row and column electrodes must be energized. Wherever an energized row and column intersect, a visible pixel will appear. To excite a pixel, a voltage potential must be applied across the LC material. For the example of Fig. 9-13, if a voltage is applied to column 638 and row 1 is connected to ground, pixel (638,1) should appear.

229

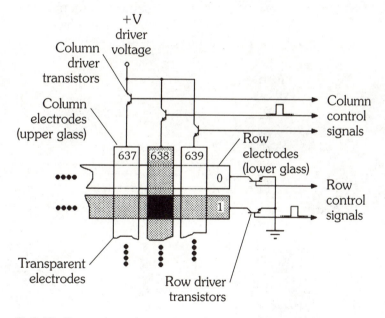

■ **9-13** *Structure and operation of a passive matrix LCD.*

A small transistor is used to switch power to each electrode. These *driver* transistors are operated by digital control signals generated in a matrix control IC which is usually located on the LCD panel itself. When a row electrode is selected, multiple column electrodes can be addressed along that row. In this way, a complete display can be developed for an entire row at a time instead of a pixel at a time. The passive matrix display is updated continually by scan-

Liquid crystal display technology

ning rows in sequence and activating each column necessary to display all of the pixels in the selected row. Most displays can update row data several times per second.

While passive matrix displays are simple and straightforward to design and build, the inherent need to scan the display slows down its overall operation. It is difficult to display computer animation or fast graphics on many passive monochrome displays. Even a mouse cursor may disappear while moving around a passive matrix LCD.

To overcome the limitations of classical passive LCDs, the *active matrix* display was developed. As shown in Fig. 9-14, each pixel is handled directly by a dedicated electrode instead of using common row and column electrodes. Individual electrodes are driven by their own transistors, so there is one transistor driver for every pixel in a monochrome display. Driver transistors are deposited onto the rear glass substrate (foundation) in much the same way that integrated circuits are fabricated. For a display with 640 × 480 resolution, a total of 307,200 (640 × 480) thin-film transistors

230

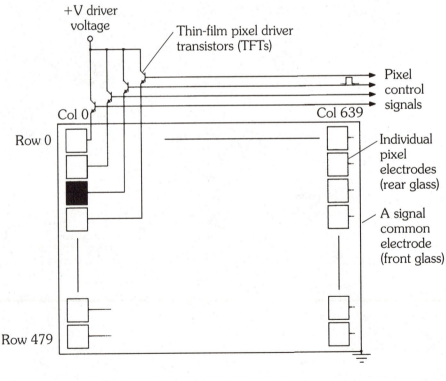

■ **9-14** *Structure and operation of an active matrix LCD (AMLCD).*

(TFTs) must be fabricated onto the rear glass. A single, huge, common electrode is deposited onto the front glass. To excite the desired pixel, it is only necessary to activate the corresponding driver transistor. The ICs that manage operation of the driver transistor array are generally included in the display panel.

When a driver transistor is fired, a potential is applied to the corresponding electrode. This potential establishes an electric field between the electrode and the common electrode on the front panel. For the example of Fig. 9-14, you will see that the pixel in row 2 and column 0 is activated simply by applying a control signal to its driver transistor. Since each pixel in an active matrix LCD can be addressed individually, there is no need to continually update the display as with passive displays. Active matrix addressing is much faster than passive matrix addressing. As a result, active matrix displays offer impressive response time with extremely good contrast. Unfortunately, active matrix LCDs are also some of the most expensive parts of your laptop or notebook computer.

Color LCD technology

The desire for high-quality flat-panel color displays continues to be somewhat of a quest for display designers. While two very effective color LCD techniques are well established, both types of color displays offer their own particular drawbacks. Passive matrix FSTN and active matrix TFT color displays are the two dominant color LCD technologies available. This section of the chapter describes today's color technologies.

Passive matrix color

Passive matrix color LCD technology is based on the operation of film-compensated super twisted nematic (FSTN or FTN) LCDs. Their principles were presented earlier in this chapter. The most striking difference between color and monochrome LCDs is that the color LCD uses three times as many electrodes as the monochrome display. This is necessary because three primary colors (red, green, and blue) are needed to form the color of each "pixel" that your eyes perceive. As shown in Fig. 9-15, each colored pixel is made up of three tiny dots.

Pixels do not actually *generate* the colors that you see. It is the white light passing through each pixel which is filtered to form the intended color. The front glass is coated with color filter material in front of each red, green, and blue dot. For example, if the dot at row 0 column 0 is supposed to be red, the green and blue dots

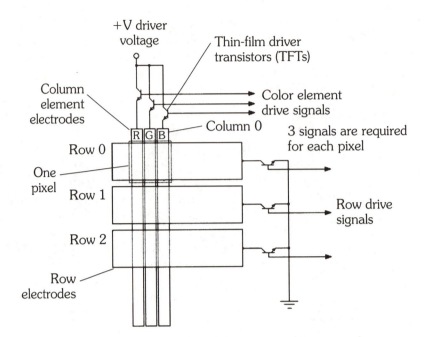

■ 9-15 *Structure of a color passive matrix LCD.*

switch at that point to block white light through all but the "red" filter. White light travels through the red filter on the front glass where it emerges as red. When the red, green, and blue dots are all on, *all* light is blocked, and the pixel appears black. If all three dots are off, all light passes through, and the pixel appears white. By controlling the three dot elements at each pixel, up to eight colors can be produced (including black and white). Intermediate color shades can be produced using color hatching schemes between adjacent dots.

The red, green, and blue (RGB) column electrodes for each pixel are deposited onto the front glass, while a single row electrode for each dot is fabricated onto the rear glass. As you might imagine, tripling the number of column electrodes complicates the manufacture of passive color displays. Not only is electrode deposition more difficult because electrodes are close together, but three times the number of column driver transistors and IC driver signals are needed. Like monochrome LCDs, the color display is updated by scanning each row sequentially and then manipulating the RGB elements for each column. Typical color LCD data can be updated several times per second.

Film-compensated super twisted nematic color displays suffer from many of the disadvantages inherent in monochrome passive matrix displays. First, response time is slow (about 250 ms). This means that no matter how fast data are delivered to the display, the image you see will only change about four times per second. Such slow update times make passive displays poor choices for fast graphic operations or animation. Their contrast ratio is a poor 7:1, which generally results in washed out or hazy displays. Viewing angles for color passive matrix LCDs are also poor at around 45°. Your clearest view of the display is when looking straight on. Constant advances in materials and fabrication techniques are improving the speed and quality of passive color displays.

Active matrix color

Active matrix color LCD technology takes the contents of monochrome active panels one step further by using three electrodes for every dot. Each electrode is completely independent and is driven by its own thin-film transistor (TFT). The three elements provide the red, green, and blue light source for each pixel that your eyes perceive. Figure 9-16 illustrates the structure of a TFT active matrix color LCD. Every electrode driver transistor and all interconnecting wiring are fabricated onto the rear glass plate. With three transistors per dot, a 640 column × 480 row color dis-

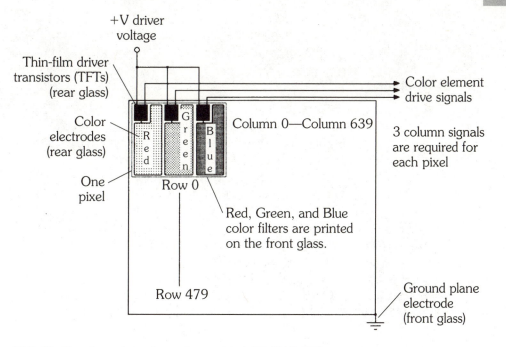

■ 9-16 *Structure of a color active matrix LCD (AMLCD).*

play requires 921,600 (640 × 480 × 3) individual transistors. Essentially, the rear plate of a TFT color display is one large 25.4-cm (10-in.) diagonal integrated circuit. The front glass plate is fabricated with a single large common electrode that every screen element can reference to.

As with passive matrix displays, the LC material used in active matrix displays does not actually generate color. The individual elements simply turn white light on or off. White light that is permitted through an element is filtered by colored material applied to corresponding locations on the front glass. When the red, green, and blue elements are all off, white light shines through the three elements, and the pixel appears white. If the red, green, and blue elements are all on, all light is blocked, and the pixel appears black. The ability to closely control contrast in individual dots allows active matrix color LCDs to produce 512 individual hues of color (256 colors are standard).

Active matrix color displays do away with many of the limitations found in passive matrix displays. Response time is very fast—on the order of 20 ms or better. Such fast response times allow the screen image to change up to 30 times per second. This provides excellent performance for graphics or animation applications. The control afforded by active matrix screens provides a brilliant contrast ratio of 60:1 with a comfortable viewing angle of 80° or so. Color active matrix displays will most likely reflect the state of the art in small-computer technology for quite some time.

Dual-scan color

Cost has been a major constraint for active matrix displays. The addition of an active matrix color display can raise the cost of a notebook or subnotebook computer by up to $1000 (U.S.). Designers have sought to improve the speed of flat-panel displays without incurring the cost of active matrix technology. The dual-scan technique is a recent improvement for passive matrix displays. Instead of scanning the entire display in one pass, the display area is broken up into upper and lower halves. Each half of the display is scanned independently. This allows any one row in the display to be updated at least twice as fast. Passive matrix performance is improved without a substantial increase in display complexity or cost. The disadvantage to dual-scan displays is that a faint horizontal bar becomes visible along the center of the display (where the upper and lower scanning areas meet). Some users may find this objectionable.

234

Plasma display technology

Gas plasma (or simply *plasma*) displays use points of ionized gas to form images. Although plasma displays are a somewhat older flat-panel display technology, you will still find many larger notebook computers offering gas plasma or LCD options. Plasma displays offer some advantages over LCDs. First, plasma displays operate in a fully transmissive viewing mode. Since ionized gas actually generates light, no backlighting is needed, and the display is easily visible in bright light or direct sunlight. Contrast is very good (usually a minimum of 50:1), and the viewing angle is at least 120°. However, plasma displays are not without their disadvantages. Because only one gas is used, the display is fixed at one color (although color gas plasma displays are in development). Although several gray-scale schemes do exist, plasma displays are limited to monochrome operation. Power supply requirements are also much more involved for plasma displays. Two high-voltage dc supplies (80 to 100 V dc and 130 to 150 V dc) are needed to drive the display. This part of the chapter introduces you to operating principles of gas plasma displays (GPDs).

Construction and operation

The typical construction of a GPD is shown in Fig. 9-17. Column electrodes are etched onto the front glass layer and then coated with a dielectric material as well as a layer of magnesium oxide (MgO). Magnesium oxide prolongs display life by preventing ion shock damage to the electrodes during gas discharge. Anode row electrodes are fabricated onto the rear glass and then coated with a dielectric material and a thin layer of magnesium oxide just like the front glass. Front and rear glass assemblies are sealed together and separated by small spacers placed at regular intervals between the layers. The discharge gap formed between the two glass layers is filled with the primary discharge gas, usually neon with a small amount of xenon. Neon produces the characteristic orange-red color that GPDs are noted for. Xenon stabilizes the neon's discharge characteristics and improves the display's light emissions.

The structure of a passive matrix GPD is illustrated in Fig. 9-18. With passive matrix designs, each row electrode is scanned in sequence. When a row is made active, each column where a pixel must be written is fired with a *writing voltage* (V_w, typically 130 to 140 V dc). Writing voltage levels will ionize the gas at desired points. Points that are already lit need only be refreshed with a *sustain voltage* (V_s, usually 80 to 100 V dc). Sustain voltage

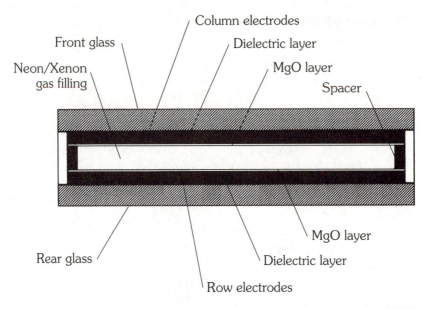

■ **9-17** *Cross-section of a conventional neon gas plasma display (GPD) assembly.*

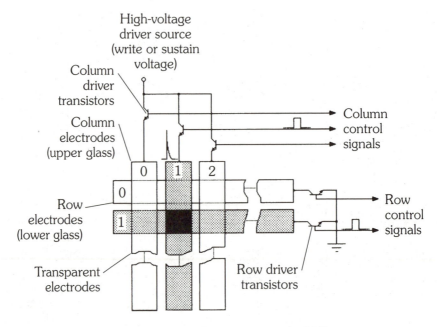

■ **9-18** *Structure and operation of a passive matrix GPD.*

maintains the gas ionization level at a reasonable brightness level. Dots that must be extinguished are left unpowered. An absence of sustaining voltage stops the ionization of gas at that point.

Troubleshooting flat-panel displays

Up to now, you have learned about LCD and plasma display panels themselves. Each type of display panel requires a substantial amount of circuitry to perform properly in a computer system under the control of a microprocessor. This part of the chapter shows you the ICs and circuitry that interface a flat-panel display (LCD or GPD) to a computer. Once you understand how flat-panel displays are driven, you will be ready to start troubleshooting.

A display system

There are seven major parts of an LCD system as illustrated in Fig. 9-19: a microprocessor, a system controller (if used), some amount of video memory (VRAM), a backlight voltage source, a highly integrated display controller IC, a contrast control, and the

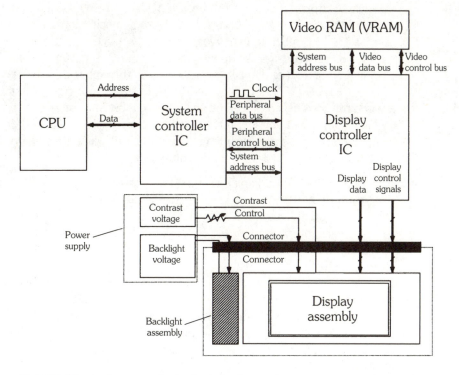

■ **9-19** *Block diagram of a typical display system.*

flat-panel display assembly itself. This is a configuration typical of notebook and subnotebook computer assemblies. If the system were designed for a GPD, the contrast control would be replaced with a brightness control, and a single ac backlight voltage would be replaced with two high-voltage dc sources.

The microprocessor is the heart of all small computers. A microprocessor is responsible for executing the instructions contained in BIOS or core memory (RAM). As the CPU executes its program(s), it directs the operations of a system controller IC. A system controller is a sophisticated ASIC which is used to handle the majority of a small computer's overhead operations. While it is not mandatory that a computer utilize a system controller, a single controller IC can effectively replace dozens of discrete logic ICs.

The display controller IC is addressed by the system controller over the common system address bus. This is the "video adapter" for small computers. Once the display controller is addressed, the system controller writes display information and commands over a secondary (or peripheral) data bus. A clock and miscellaneous control signals manage the flow of data into the display controller. Display data are interpreted and stored in video RAM (VRAM). Each pixel in the physical display can be traced back to a specific logical location (address) in VRAM. As new data are written to the display controller, VRAM addresses are updated to reflect any new information. During an update, the display controller reads through the contents of VRAM and sends the data along to the flat-panel display. There are two other signals required by an LCD assembly: contrast voltage and backlight input voltage. If the display were gas plasma, it would require a brightness control voltage, a high-voltage dc write voltage (V_w), and a high-voltage dc sustain voltage (V_s).

A complete flat-panel display assembly is shown in Fig. 9-20. The plastic outer housings (marked K1 and K22) form the cosmetic shell of the display panel. An LC cell with its driver ICs and transistors (marked E6) is mounted to the front housing. A number of insulators may be added to protect the LC cell from accidental short circuits. Note that the front polarizer layer is built into the LC cell. A rear polarizer (marked K15) is placed directly behind the LC cell, followed by the translucent backlight diffuser panel (marked K16). The backlight mechanism is a long, thin CCFT (marked K13) located along the bottom of the diffuser. Several spacers-brackets are used to secure the diffuser panel and tie the entire assembly together. Finally, a rear housing is snapped into place to cover the display. A cable (marked K19) connects the display assembly to the motherboard.

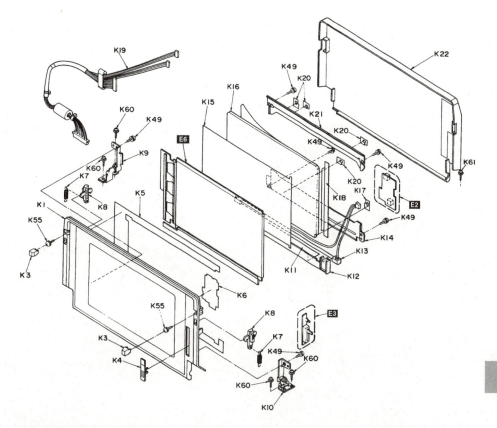

■ **9-20** *Exploded diagram of a display system. (Courtesy of Tandy Corporation)*

Symptoms and solutions

Symptom 1 *One or more pixels are defective. The defective pixel may be black (opaque), white (clear), or fixed at some color.* Before beginning any repair procedure, turn off your computer and initialize it from a cold start. A cold start ensures that your pixels are not being locked up due to any possible software glitch. If the questionable pixels disappear, there may be a bug in your application software, not in your hardware.

This is a symptom that commonly occurs with active matrix display panels. Since each screen dot is driven by either one or three individual driver transistors, the loss of one or more drivers will ruin a pixel. For monochrome displays, the single driver transistor may be open (pixel will not turn on) or shorted (pixel will not turn off). For color displays, damage to a driver transistor may cause a certain color to appear and remain fixed on the screen as long as the computer is running. Unfortunately, there is no way to repair a failed screen driver transistor. Active matrix flat-panel circuitry is fabricated much the same way as any integrated circuit, so when any part of the IC fails, the entire IC must be replaced. Your best course is simply to replace the suspect LCD assembly.

If a new flat-panel display fails to correct the trouble, replace the VRAM IC(s). One or more address locations may be defective, resulting in faulty video data being provided to the display. It is unusual for a problem in VRAM to manifest itself in this way, but you should be prepared for this possibility. If you do not have the proper tools or inclination to perform surface-mount soldering, you may prefer to replace the entire motherboard.

Symptom 2 *There is poor visibility in the LCD. The image is easily washed out in direct or ambient light.* The vast majority of small-computer LCDs operate in the *transmissive* mode—the light that your eyes see is generated from a backlight system. As a result, the strength and quality of the backlight directly affect the display's visibility. If the computer is older and has accumulated a great deal of running time, the EL backlight panel or CCFT(s) may be worn out or failing. The high-voltage power supply that is operating the backlight may also be faulty.

Check your LCD's contrast control. Contrast adjusts the amount of driver voltage that is used to straighten the LCD material. Less driver voltage straightens the liquid crystal less. This results in less contrast. If contrast is already high or maximum, disassemble the display to expose your contrast adjustment. Measure contrast voltage across the adjustment. Your multimeter should read the entire contrast display voltage. If contrast voltage is low or absent, troubleshoot or replace your dc-dc converter. Measure contrast voltage from the center (adjustable) leg of the control with respect to ground and vary the adjustment. If voltage is zero or remains fixed while the adjustment is varied, replace the defective contrast control or module.

Disassemble your flat-panel assembly to expose the backlight unit (either an EL panel or a CCFT assembly) as well as the inverter power supply. *Gently* brush away any dust or debris that may have accumulated on the EL panel, CCFT(s), or diffuser. If the light source has been badly fouled by dust, retry the system with the cleaned light source. Use your multimeter to measure the ac output voltage from your inverter power supply. A working inverter should output 150 to 200 V ac for an EL panel or 200 to 1000 V ac for CCFTs. Remember to use **extreme** caution when measuring high voltages. If your inverter output is low or nonexistent, troubleshoot or replace your faulty backlight power supply. If the inverter output voltage appears normal, replace the defective backlight assembly.

Symptom 3 *There is poor visibility in the GPD. The image is not bright.* Gas plasma displays generate their own light as a natural by-product of ionized gas. The brightness shown by the display is then heavily dependent on the voltage levels used to ionize (write) and light (sustain) the gas. By adjusting the sustain voltage level, the display's brightness should vary. Check your plasma display's brightness control. If brightness is already set high or at maximum, disassemble your display to expose your brightness adjustment. Use your multimeter to measure the brightness control voltage across the adjustment. Remember to use **extreme** caution when measuring high voltages. If the brightness control voltage is low or absent, check your plasma write and sustain power supplies. Measure the brightness control voltage at the center of the adjustment versus ground and vary the adjustment. If voltage is zero or remains fixed while the adjustment is being varied, replace the faulty brightness control or module.

Measure your plasma drive voltages. You should measure the sustain voltage (V_s) and the write voltage (V_w). Sustain voltage should measure 80 to 100 V dc, and write voltage should measure 130 to 150 V dc. Remember to use **extreme** caution when measuring high voltages. Some plasma displays use ac drive voltages, so consult your owner's or service manual for the particular display specifications. If you are unable to correct a low supply voltage with output adjustments on the supply itself, or if the supply outputs measure very low, repair or replace your plasma power supply.

Symptom 4 *The display is completely dark. There is no apparent display activity.* This symptom assumes that your computer has plenty of power and attempts to boot up with all normal disk activity. You simply have no display. If there are no active LEDs to indicate power or disk activity, there may be a more serious problem with your system. Begin by removing all power from your system. Remove the outer housings of your computer and inspect all connectors and wiring between the motherboard and display. Tighten any loose connectors and reattach any loose or broken wiring. Defective connections can easily disable your display.

Reapply power to your computer and check the display voltage(s) powering your display system. Typical LCDs use +5 V dc (or +3.0 or +3.3 V dc for low-voltage systems) and ±12 V dc. Plasma displays use high dc voltages around 80 to 100 V dc and 130 to 150 V dc.

Remember to use **extreme** caution when measuring high voltages. If your display is not receiving the appropriate voltages, check your power supply and/or dc-dc converter(s). Replace any defective power components.

When your display is connected securely and is receiving the appropriate voltage levels, you should suspect a fault in the display controller IC or somewhere within the display assembly itself. Use your logic probe to check the data lines and control signals from the display controller to the display assembly. If you have service charts available for the controller, you should have no difficulty locating the appropriate signals. You should expect to see high-frequency pulse signals on each of the data and control lines since the display must be updated continuously. If you see one or more data lines frozen in a logic 1 or logic 0 state, this may indicate a defective display controller. A faulty logic state or any of the display controller's outputs also suggests a faulty display controller IC. In either case, the controller should be replaced. If you do not have the tools or inclination to perform surface-mount soldering, you should probably replace the entire notebook or laptop motherboard. If you find the signals from the display controller to be intact, the circuitry within the display assembly itself is probably faulty. Your best course would then be to replace the entire display assembly.

Symptom 5 *The display appears erratic. It displays disassociated characters and garbage.* This is another symptom that assumes your small computer has plenty of power and attempts to boot up with normal disk activity. Your display is simply acting erratically. If no power indicators or disk activity LEDs are lit, there may be a much more serious problem in your system. Remove all power from your system and remove the outer housings to expose the motherboard and display assembly. Inspect all cables and connectors between the motherboard and display assembly. Tighten any loose connectors and secure any loose or broken wiring. Defective connectors or wiring can easily interfere with normal display operation. You may also wish to check the voltage levels powering your display.

When all connections are intact, you should suspect a fault in the display controller or video memory (VRAM). Unfortunately, the great volume of data flowing from the system controller to the display controller, from VRAM to the display controller, and from the display controller to the display itself makes a comprehensive test virtually impossible without sophisticated test equipment. On a symptomatic basis, you should replace the display controller first because it is at the crux of the display system. If a new display controller IC does not correct the problem, try replacing the VRAM IC(s). Defective memory can cause erratic display performance. If you are unable to perform surface-mount soldering, you should simply replace the laptop or notebook motherboard. A motherboard change will exchange the display controller and VRAM together. If the problem persists, try replacing the display assembly itself.

243

10

Using MONITORS
for diagnostics

CHAPTER 5 TOOK YOU THROUGH A COMPLETE TUTORIAL ON testing and aligning a CRT-based monitor. To successfully test and align a monitor (Fig. 10-1), it is necessary to display a comprehensive set of test patterns and ensure that each image follows a predictable set of criteria or behaviors. Traditionally, specialized hardware tools have been employed as test pattern generators (e.g., the line of MONTEST signal generators illustrated in Chapter 3). While signal generators produce clean, precise patterns, the outright cost for such a device is often more than the casual user or electronics enthusiast can justify. Even professional service shops are looking for cost-saving options. Since homes and businesses with monitors also have PCs, it makes good sense to use the existing *PC* as a test instrument. This chapter is intended to introduce you to MONITORS, which is the companion DOS utility disk for this book. You will learn how to obtain your own copy of MONITORS, install it on your PC, and use it productively in a matter of minutes.

Before we go any farther, you should understand that you *do not* have to purchase the companion disk in order to troubleshoot a monitor effectively—that would certainly not be fair to you. The majority of this book is written without relying on the use of a companion disk. For anyone who wishes to perform alignments and test their video boards, however, you will find MONITORS to be an inexpensive but handy addition to your toolbox.

All about MONITORS 2.01

MONITORS is a self-contained video board diagnostic and test pattern generator. It is designed to work with MDA, CGA, EGA, VGA, and SVGA video systems and is compatible across a wide se-

244

■ **10-1** *An NEC MultiSync 4PG monitor. (NEC Technologies, Inc.)*

lection of hardware (both video adapters and monitors). MONITORS allows you to check a PC's video adapter capabilities and perform a suite of standard test-alignment procedures after a repair is complete. Since the first edition of this book, MONITORS has been updated to include burn-in features (previously a standalone utility) as well as streamlined user input and performance. Once the monitor is aligned, the burn-in feature will simply work the video system for several hours (or several days) to ensure that there is no supplemental work that needs to be done. The system requirements for MONITORS are listed in Table 10-1.

■ **Table 10-1 System requirements for MONITORS 2.01**

CPU:	Intel i286, i386, i486, or Pentium
DOS:	ver 5.0 or later (works under Windows 3.1 and Windows 95 in the DOS mode)
RAM:	640 kbytes (conventional memory only)
Mouse or Trackball:	not required
Video System:	MDA, CGA, EGA, VGA, SVGA
Hard Drive Space:	about 0.5 Mbyte for hard drive installation (optional)

Obtaining your copy of MONITORS

A copy of MONITORS can be obtained directly from Dynamic Learning Systems. An advertisement and order form are included at the end of this book. Feel free to photocopy the order form (or tear out the page), fill out the required information, and enclose a check or money order for the proper amount. Please remember that all purchases must be made in U.S. funds. When filling out the order form, you may select the companion disk alone, a 1-year subscription to our premier newsletter *The PC Toolbox*™, or take advantage of a very special rate for the disk *and* subscription combined. Subscribers also receive extended access to the Dynamic Learning Systems BBS which allows you to exchange e-mail with other PC enthusiasts and download hundreds of DOS and Windows PC utilities. Be sure to specify your desired disk size (3.5 or 5.25 in.).

Installing and starting from the floppy drive

When the companion disk arrives in the mail, your first task should be to make a backup copy on a blank floppy disk. You can use the DOS `diskcopy` function to make your backup. For example, the command line:

```
C:\> diskcopy a: a: <ENTER>
```

will copy the original disk. Keep in mind that you will have to do a bit of disk swapping with this command. If you wish to use a floppy drive besides `a:`, you should substitute the corresponding letter for that drive. If you are uncomfortable with the `diskcopy` command, refer to your DOS manual for additional information. MONITORS is designed to be run directly from the original floppy disk, so you can keep the original disk locked away while you run from the copy. This allows you to take the disk from machine to machine so that you will not clutter your hard drive or violate the license agreement by loading the software onto more than one machine simultaneously.

1. To start MONITORS from the floppy drive, insert the floppy into the drive and type the letter of that drive at the command prompt and press enter. The new drive letter should now be visible. For example you can switch to the `a:` drive by typing:

    ```
    C:\> a: <ENTER>
    ```

    The system will respond with the new drive letter:

    ```
    A:\>_
    ```

2. Then type the name of the utility:

```
A:\> monitors <ENTER>
```

3. The program will start in a few moments, and you will see the main menu. If you are using a floppy drive other than `a:`, you should substitute that drive letter (such as `b:`) in place of the `a:`.

Installing and starting from the hard drive

If you will only be using one PC and have an extra 0.5 Mbyte or so, you should still go ahead and make a backup copy of the companion disk as described in the previous section, but it would probably be more convenient to install the utilities to your hard drive. There is no automated installation procedure to do this, but the steps are very straightforward.

1. Boot your PC from the hard drive and, when you see the command prompt, switch to the root directory by typing the `cd\` command:

```
C:\> cd\ <ENTER>
```

The system should respond with the root command prompt:

```
C:\>_
```

2. Use the DOS `md` command to create a new subdirectory that will contain the companion disk's files. One suggestion is to use the name MONITORS such as:

```
C:\> md monitors <ENTER>
```

Then switch to the new subdirectory using the `cd\` command:

```
C:\> cd\monitors <ENTER>
```

The system should respond with the new subdirectory label such as:

```
C:\MONITORS>_
```

You may of course use any DOS-valid name for the subdirectory or nest the directory under other directories if you wish. If you are working with a hard drive other than `c:`, substitute that drive label for `c:`.

3. Insert the backup floppy disk into the floppy drive. Use the DOS copy command to copy all of the floppy disk files to the hard drive such as:

```
C:\MONITORS> copy a:*.* c: <ENTER>
```

This instructs the system to copy all files from the `a:` drive to the current directory of the `c:` drive. Since MONITORS is not distributed in compressed form, uncompression (or un-zipping) is not needed.

4. After all files have been copied, remove the floppy disk and store it in a safe place. Then, type the name of the program:

`C:\> monitors <ENTER>`

The startup screen should appear almost immediately.

Using MONITORS

When MONITORS is started, you will see a brief initialization text followed by a disclaimer and media warranty (press any key to clear the disclaimer). You will then see the main menu (Fig. 10-2). The top portion of the main menu lists the utility name and version. Sixteen selections are available from the main menu (the main menu for version 2.01 is shown, but other versions may appear slightly different). The first two menu selections allow you to test the video adapter and set the appropriate video mode for the monitor in use. Chapter 5 explains how to apply each test pattern in detail. The burn-in feature allows you to run your monitor unattended for prolonged periods. If you are done with the program, press <ESC> at the main menu to return to DOS.

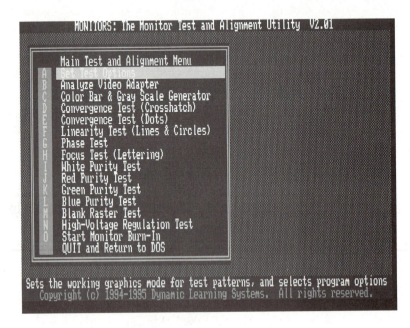

■ **10-2** *The MONITORS 2.01 main menu.*

Testing your video adapter

MONITORS allows you to test the capabilities of your video adapter by selecting option B from the main menu. The General Adapter Specifications shown in Fig. 10-3 illustrate an overview of the adapter's capabilities. Six key pieces of information are provided. The *Chipset in use* identifies the particular video chipset employed by the adapter. MONITORS is designed to identify up to 16 types of video chipsets (a listing is provided in Table 10-2). If the chipset cannot be identified, MONITORS will indicate an "Unknown Chipset." In addition to the type of chipset in use, the *Chipset revision* is also identified wherever possible. If the chipset type is not known or the revision cannot be determined, MONITORS will indicate an "Unknown Revision."

The total amount of *Video Memory* available on the video board is listed next. The specification is given in KB. For VESA-compatible video boards, the *VESA Version* will be shown. Otherwise, a "No VESA Driver Installed" message is shown. If I/O registers are used with the video board, the *I/O Register Base* address will be listed. Otherwise, an "I/O Registers Not Installed" message is displayed. When a video board with a built-in VESA BIOS is used, the *VESA*

MONITORS: The Monitor Test and Alignment Utility V2.01

Video Adapter Analysis: GENERAL SPECIFICATIONS

Chipset in use......: ATI VGA Wonder Chipset
Chipset revision....: ATI 68800
Video Memory........: 2048 KB
VESA Version........: No VESA Driver Installed
I/O Register Base...: IO Registers Not Used
VESA BIOS Status....: Not Installed

General Video Adapter: Video Graphics Array
Set Test Display Mode: VGA Color Display: 640x480x16

Video Board Specifications Finished...Press <ENTER> for next page...
Copyright (c) 1994-1995 Dynamic Learning Systems. All rights reserved.

■ **10-3** *The first video adapter analysis report page (Board Characteristics).*

■ Table 10-2 MONITORS compatibility in identifying video chipsets

1. Ahead Systems
2. ATI
3. Avance Logic
4. Cirrus Logic
5. Compaq QVision
6. IBM XGA
7. NCR
8. Oak Technologies
9. Paradise
10. S3
11. Trident
12. Tseng ET3000
13. Tseng ET4000
14. VESA (compatible)
15. Video 7
16. Weitek

BIOS Status is listed last. When VESA BIOS support is not included, a "Not Installed" message is shown. The last two entries indicate the *General Video Adapter* (the generic video system in use) and the recommendation for *Set Test Display Mode* (video mode that should be used for testing and alignment).

The next step in adapter analysis identifies each major video mode that the adapter can successfully emulate. A typical listing is shown in Fig. 10-4. Available modes are highlighted in green, while incompatible modes are highlighted in red (although Fig. 10-4 does not show such color differences clearly). Note the large number of VESA modes that are tested. One important thing to keep in mind when reviewing Video Mode Compatibility is that the available modes do not necessarily mean that the attached monitor is *capable* of working in every mode. Once you have finished reviewing the available modes, press the <ENTER> key to return to the main menu.

Choosing video test modes

By default, test patterns are generated in VGA at a resolution of $640 \times 480 \times 16$. This automatically accommodates the broad base of VGA and SVGA video systems in service today. Option A from the main menu allows you to override this default and select a dif-

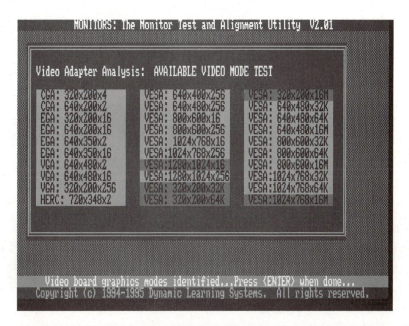

■ **10-4** *The second video adapter analysis report page (Available Modes).*

ferent video mode for test pattern displays as shown in Fig. 10-5. There are six selections: MDA (for text-only displays), CGA, EGA, VGA, SVGA, and an Auto-Select mode. If you are not certain what mode is appropriate for your system, the Auto-Select selection will generally make a reasonable choice for you. You can also switch back and forth between different screen modes at your discretion. Keep in mind that not all test patterns are available in all screen modes. Table 10-3 shows the availability of test patterns for each screen mode.

You will also notice two additional selections marked A and B. The first selection is "MONITORS Help Mode." Normally, the program will display a brief text description of the test before starting it. If you toggle the help mode off, no text information will be displayed, and the test pattern will simply start. Novices may find the help mode useful, so it is on by default. The next selection is the "Menu Blink Mode." This feature was intended to assist visually impaired users by causing the highlighted selection in the main menu to blink. This feature is on by default, but if you find it annoying or otherwise objectionable, just toggle it off. Once you are done selecting a screen mode or features, press the <ESC> key to return to the main menu.

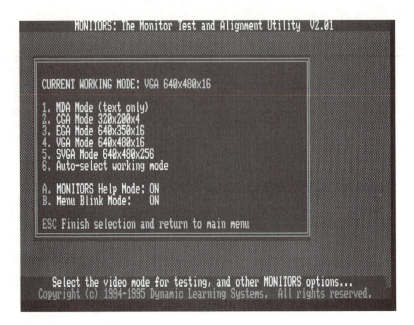

■ **10-5** *The MONITORS 2.01 Test Options menu.*

Color bars

The *color bar* display is available by pressing option C from the main menu. In the SVGA mode, a palette of 256 colors is displayed (assuming full SVGA compatibility). In VGA and EGA, 16 colors are shown as in the VGA display of Fig. 10-6. The CGA mode shows 4 colors. As you might expect, color bars are unavailable in the MDA mode. Color bars are not used for specific alignment tests, but rather for general quality testing and settings of brightness and contrast. Color bars provide a good overview of monitor quality.

When you are through with the color bars, press any key on the keyboard. Displays in the SVGA and VGA modes will switch from color bars to a set of 10 gray bars as shown in Fig. 10-7. Displays in the EGA and CGA modes will return to the main menu. This is a *gray scale linearity* test which should produce 10 even graduations of gray (including black and white). As with color bars, gray scale linearity is used to judge the quality of a monitor's gray scale generation. Press any key on the keyboard to return to the main menu.

Convergence test (crosshatch)

Static convergence is an important test for color monitors of all resolutions. When option D is selected from the main menu, a white

Pattern	SVGA	VGA	EGA	CGA	MDA
Color bars	256 col	16 col	16 col	4 col	n/a
(Gray scale)	10 gry	10 gry	n/a	n/a	n/a
Convergence (crosshatch)	graphic	graphic	graphic	graphic	n/a
Convergence (dots)	graphic	graphic	graphic	graphic	n/a
Linearity	graphic	graphic	graphic	graphic	ASCII
Phase	graphic	graphic	graphic	graphic	ASCII
Focus	ASCII	ASCII	ASCII	ASCII	ASCII
White purity	graphic	graphic	graphic	graphic	n/a
Red purity	graphic	graphic	graphic	n/a	n/a
Green purity	graphic	graphic	graphic	n/a	n/a
Blue purity	graphic	graphic	graphic	n/a	n/a
Blank raster	graphic	graphic	graphic	graphic	ASCII
High voltage	graphic	graphic	graphic	graphic	ASCII

wh = white
n/a = not available
col = colors
gry = gray

253

grid will appear as shown in Fig. 10-8. The use of white is important since all three electron guns are running at equal amplitude. Other colors may be more visually appealing, but only pure white provides useful information. When you observe the convergence grid, there should be no other colors (blue, green, or red) bleeding out from around the grid edges. If there are, you will need to perform a static convergence alignment as described in Chapter 5. By press-

■ **10-6** *The basic color bar display.*

■ **10-7** *The gray scale linearity display.*

ing the M key, the grid color turns magenta (red and blue electron guns *only*). This allows you to adjust the magenta convergence magnets. By pressing the W key, the grid returns to white, and you can adjust the white convergence magnets. You can switch back and forth between magenta and white as needed by alternately pressing the M and W keys. Pressing any other key will return to the main menu. A convergence crosshatch pattern is available in all screen modes except MDA.

Convergence test (dots)

This is another test for *static convergence.* When option E is selected from the main menu, a pattern of white dots will appear as shown in Fig. 10-9. The use of white dots remains important because all three electron guns must be running at equal amplitude. As with the crosshatch pattern, other colors may be more visually appealing, but only pure white provides useful information. When you observe the convergence dots, there should be no other colors

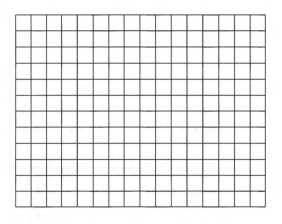

■ **10-8** *The convergence (crosshatch) display.*

(blue, green, or red) bleeding out from around the individual dots. If there are, you will need to perform a static convergence alignment as described in Chapter 5. By pressing the M key, the dot color turns magenta (red and blue electron guns only). This allows you to adjust the magenta convergence magnets. By pressing the W key, the dots return to white, and you can adjust the white convergence magnets. You can switch back and forth between magenta and white as needed by alternately pressing the M and W keys. Pressing any other key will return to the main menu. The convergence dot pattern is available in all screen modes except MDA.

Linearity test

Option F from the main menu selects the *linearity test* pattern as shown in Fig. 10-10. By using a series of geometric shapes (i.e., circles and squares), linearity tests show the "evenness" of both horizontal and vertical raster. If any portion of the screen image appears compressed or expanded, linearity should be adjusted. When distortion occurs in the horizontal orientation, horizontal linearity needs to be adjusted. When distortion occurs in the vertical direction, vertical linearity needs to be adjusted. Graphic lin-

■ **10-9** *The convergence (dots) display.*

earity patterns are available in all screen modes except MDA. An ASCII pattern is used to check for MDA display linearity. Press any key on the keyboard to return to the main menu.

Phase test

Option G from the main menu selects the *phase test* pattern as shown in Fig. 10-11. Ideally, an image should be horizontally centered in the raster. You can see the phase of an image by turning up brightness until the raster is visible. The thin white box around the perimeter of the image allows you to adjust horizontal phase until the image is reasonably centered. This allows horizontal width to be adjusted evenly. If phase is incorrect, widening the image will cause the edge closest to the edge of the raster to "run off" the display. Graphic phase test patterns are available in all screen modes except MDA. An ASCII pattern is used for the MDA mode. Press any key on the keyboard to return to the main menu.

Focus test

The *focus test* pattern is displayed by selecting option H from the main menu. One of the best tests of focus is in the display of ASCII

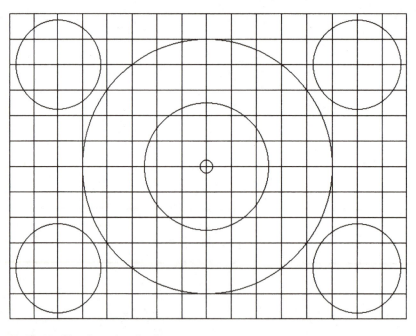

■ **10-10** *The linearity display.*

■ **10-11** *The screen phase adjustment display.*

test, so Fig. 10-12 illustrates the typical focus pattern used for *all* screen modes. When the text image appears out of focus, high-voltage levels to the CRT's focus grid(s) can be adjusted to compensate. Older monitors provide a panel-mounted focus control, but newer monitor designs place the focus control on the monitor's main PC board. When testing and adjustment are complete, press any key on the keyboard to return to the main menu.

The purity tests

Options I, J, K, and L are color purity selections used to check color gun consistency across the display. The white purity selection (I) is by far the most common and widely used mode. By filling the screen with a pure white image, it is possible to check for discoloration or uneven coloring that may indicate the need for manual degaussing. The other color selections may also be selected at your discretion. When you are done reviewing a purity test pattern, press any key on the keyboard to return to the main menu. Note that the white purity test is available in all screen modes but MDA. However, the red, green, and blue purity tests are not available in the CGA and MDA modes.

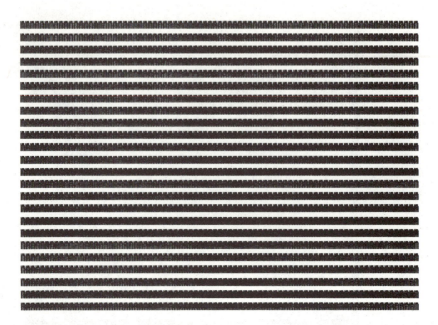

■ **10-12** *The focus adjustment display.*

Blank raster test

Option M from the main menu starts the *blank raster test.* Essentially, the blank raster is just a black image. This allows you to increase screen brightness and check the presence of raster without the distraction of a screen image. Blank raster is available in all screen modes. When you are through, press any key on the keyboard to return to the main menu.

High-voltage test

Option N from the main menu selects the *high-voltage test* as illustrated in Fig. 10-13. This pattern is used to check high-voltage regulation by blinking the middle white box on and off. Ideally, the boarder should not flinch significantly. If it does, you may need to make repairs to the high-voltage system. Once the test is complete, press any key on the keyboard to return to the main menu. A graphic image is used to test high-voltage regulation in all screen modes but MDA. An ASCII image is used in the MDA mode.

Monitor burn-in

Once a repair is finished, it is often helpful to let the repaired monitor run for several hours (or even several days) to see that the re-

■ **10-13** *The high-voltage regulator display.*

pair will hold. Unfortunately, when the same image is allowed to dwell for long periods of time at high brightness, there is a small but persistent risk of CRT phosphor burn. Although phosphor burn is almost unheard of today, it is still a wise policy to periodically alternate the screen image. MONITORS offers a burn-in mode (option O from the main menu) which displays a continuously repeating series of test patterns. After the cycle starts, you can walk away from the monitor indefinitely. To stop the cycle and return to the main menu, press the <ESC> key.

When trouble occurs

The system requirements for MONITORS are very mild compared to other commercial software in the marketplace today. As a result, the probability of problems with the software should be extremely rare. Make sure that the monitor is connected properly and that it is appropriate for the video board installed in the PC.

Check the screen mode

The most common error is choosing a screen mode that is incompatible with the monitor. When choosing a screen mode, make sure that the monitor is capable of supporting it. For example, choosing an SVGA screen mode for an EGA monitor will result in display problems. However, this should not cause an error in the software. Follow the proper key sequence to return to the main menu and select a more appropriate screen mode.

Not all video boards support the same video mode in the same way, so the SVGA mode may not work properly for all video systems. If the displays in the SVGA mode appear distorted, return to the main menu and select the VGA mode. Virtually all SVGA video boards and SVGA monitors support VGA. Other than the initial color bar display, the VGA and SVGA test patterns are identical.

When problems persist

If problems persist, write your symptoms and system setup on a sheet of paper along with your name, address, and telephone and fax numbers (if possible) and mail or fax it to:

Customer Service
Dynamic Learning Systems
P.O. Box 282
Jefferson, MA 01522-0282 USA
Fax: 508-829-6819
BBS: 508-829-6706
CompuServe: 73652,3205
Internet: sbigelow@cerfnet.com
WWW: http://www.dlspubs.com/

Making MONITORS better

Dynamic Learning Systems is committed to providing useful and reliable software tools, so we welcome your comments and suggestions on how we can improve MONITORS and BURNIN. You can also reach us on-line through our BBS or on CompuServe. Feel free to send your feedback to:

Customer Service
Dynamic Learning Systems
P.O. Box 282
Jefferson, MA 01522-0282 USA
Fax: 508-829-6819
BBS: 508-829-6706
CompuServe: 73652,3205

Monitor troubleshooting index

A

Chapter 2: CRT monitor basics

Symptom 1 *The screen images are dim or dark.*
- *a.* check brightness and contrast controls
- *b.* check for the presence of a glare shield
- *c.* check high voltage (adjust if necessary)
- *d.* check B+ voltage (adjust if necessary)
- *e.* troubleshoot or replace B+ supply if necessary
- *f.* check other power supply outputs
- *g.* troubleshoot or replace power supply if necessary
- *h.* try rejuvenating the CRT
- *i.* replace the CRT

261

Symptom 2 *The image contains dark blacks and overdriven whites.*
- *a.* check the power supply outputs
- *b.* troubleshoot or replace the power supply if necessary
- *c.* try rejuvenating the CRT
- *d.* replace the CRT

Symptom 3 *The monitor displays poor colors or a nonlinear gray scale.*
- *a.* check video signal at each amplifier
- *b.* adjust video signal levels if necessary
- *c.* try replacing the video amplifier board
- *d.* replace the CRT

Symptom 4 *The raster appears unusually bright and may appear colored.*
- *a.* try rejuvenating the CRT
- *b.* replace the video amplifier board
- *c.* replace the CRT

Chapter 4: video adapters

Symptom 1 *The computer is on, but there is no display.*
   a. check the monitor installation
   b. check the video adapter installation
   c. check the adapter's DIP switch-jumper configuration
   d. check the adapter for hardware conflicts
   e. check for software driver problems or conflicts
   f. check for system memory conflicts
   g. replace the video adapter

Symptom 2 *There is no display (the computer sounds a series of beeps).*
   a. verify that the beep code corresponds to a video problem
   b. check the video adapter installation
   c. replace the video adapter
   d. troubleshoot/replace the motherboard

Symptom 3 *You see large black bands at the top and bottom of the display in some screen modes, but not in others.*
   a. make sure the adapter is appropriate for the monitor being used
   b. check the monitor for autosizing
   c. adjust/troubleshoot the monitor
   d. replace the video adapter

Symptom 4 *The display image rolls.*
   a. make sure the adapter is appropriate for the monitor being used
   b. replace the video adapter
   c. adjust/troubleshoot the monitor

Symptom 5 *An error message appears on system startup indicating an invalid system configuration.*
   a. check system CMOS (replace backup battery if necessary)
   b. check the video adapter configuration
   c. check/troubleshoot/replace the motherboard

Symptom 6 *Garbage appears on the screen or the system hangs up.*
   a. make sure the video adapter is appropriate for the monitor being used
   b. select an alternate video mode
   c. try toggling the monitor off and on
   d. check for hardware conflicts between the adapter and other expansion boards
   e. check for software conflicts between drivers or TSRs

Chapter 6: power supply troubleshooting

Linear supplies

Symptom 1 *The monitor is dead. There is no raster and no picture.*

 a. check the monitor's installation
 b. check the ac input to the monitor
 c. check the monitor's fuse(s)
 d. check for loose connectors/wiring or damage to the power PC board
 e. check/replace regulator(s)
 f. check/replace filter(s)
 g. check/replace rectifier(s)
 h. check/replace power transformer
 i. replace power supply or main PC board

Symptom 2 *The monitor operates only intermittently. You might see the raster and power indicator LED blink on and off.*

 a. check the monitor's installation
 b. check the ac input to the monitor
 c. check for loose connectors/wiring or damage to the power PC board
 d. check for thermal intermittent problems
 e. replace the power supply or raster board

Symptom 3 *The main ac fuse fails, and the new replacement fuse also fails.*

 a. isolate the power supply or raster board if possible
 b. inspect power supply components
 c. replace any defective (typically shorted) components
 d. replace the power supply (or the entire raster board)
 e. replace the video amplifier board if necessary

Symptom 4 *The main ac fuse fails when the power supply is cold.*

 a. check/replace power supply thermistors or posistors
 b. check/troubleshoot the degaussing circuit

Switching supplies

Symptom 1 *The monitor is completely dead. There is no raster and no picture.*

 a. check the monitor's installation
 b. check the ac input to the monitor
 c. check the monitor's fuse(s)
 d. check for loose connectors/wiring or damage to the power PC board

e. check/replace the line filter(s)
f. check/replace the input transformer
g. check/replace the rectifier(s)
h. check/replace the switching controller IC
i. check/replace the switching transistor(s)
j. check/replace the individual output rectifier(s) and regulator(s)
k. replace the power supply or main PC board

Symptom 2 *The monitor operates only intermittently. You might see the raster and power indicator LED blink on and off.*
a. check the monitor's installation
b. check the ac input to the monitor
c. check for loose connectors/wiring or damage to the power PC board
d. check for thermal intermittent problems
e. replace the power supply or main PC board

Symptom 3 *The main ac fuse fails, and the new replacement fuse also fails.*
a. isolate the power supply or raster board if possible
b. inspect power supply components
c. replace any defective (typically shorted) components
d. replace the power supply (or the entire raster board)
d. replace the video amplifier board if necessary

Symptom 4 *The main ac fuse fails when the power supply is cold.*
a. check/replace power supply thermistors or posistors
b. check/troubleshoot the degaussing circuit

Backlight supplies

Symptom 1 *The backlight appears inoperative. The LCD may seem washed out or invisible in low light.*
a. check the backlight time-out setting
b. check the inverter's dc input
c. check the inverter's ac output
d. check/replace the backlight device (CCFT or EL)
e. check/replace the backlight supply module.

High-voltage supplies

Symptom 1 *There is no raster and no picture, or there is a vertical line against the raster.*
a. check the monitor's contrast and brightness controls
b. check the monitor's installation

264

<ol type="c" start="3">
<li>check/replace the FBT</li>
<li>check/rejuvenate/replace the CRT</li>
<li>check/replace the horizontal output transistor(s)</li>
<li>check/replace the horizontal output transformer</li>
<li>check/replace the horizontal oscillator IC</li>
<li>replace the main PC board.</li>
</ol>

Chapter 7: monochrome monitor troubleshooting

Symptom 1 *Raster is present, but there is no image.*
<ol type="a">
<li>check the monitor installation</li>
<li>check for an interruption in the video drive circuit</li>
<li>check/rejuvenate/replace the CRT</li>
<li>check/replace the video drive PC board</li>
</ol>

Symptom 2 *A single horizontal line appears in the middle of the display.*
<ol type="a">
<li>check/replace the vertical oscillator</li>
<li>check/replace faulty vertical output driver transistor(s)</li>
<li>replace the main PC board</li>
</ol>

Symptom 3 *Only the upper or lower half of an image appears.*
<ol type="a">
<li>check/replace faulty vertical output driver transistor(s)</li>
<li>replace the main PC board</li>
</ol>

Symptom 4 *A single vertical line appears along the middle of the display.*
<ol type="a">
<li>check/replace the horizontal deflection yoke assembly</li>
<li>check/replace the horizontal oscillator</li>
<li>check/replace the horizontal switching transistor(s)</li>
<li>check/replace the FBT</li>
<li>replace the main PC board</li>
</ol>

Symptom 5 *There is no image and no raster.*
<ol type="a">
<li>check the monitor's installation</li>
<li>check the contrast and brightness controls</li>
<li>check/troubleshoot/replace the power supply or main PC board</li>
<li>check/replace the horizontal oscillator</li>
<li>check/replace the horizontal output transistor(s)</li>
<li>check/rejuvenate/replace the CRT</li>
<li>check/replace the FBT</li>
<li>replace the main PC board</li>
</ol>

Symptom 6 *The image is too compressed or too expanded.*
<ol type="a">
<li>check the front-panel vertical size control(s)</li>
<li>check/replace the vertical oscillator</li>
</ol>

265

    c. check for PC problems on the main PC board
    d. replace the main PC board

Symptom 7 *The displayed characters appear to be distorted.*
    a. check the monitor's location and installation
    b. check/correct the image's linearity
    c. check/correct the image's focus
    d. check/replace the FBT
    e. replace the main PC board

Symptom 8 *The display appears wavy.*
    a. check/troubleshoot/replace the power supply or main PC board

Symptom 9 *The display is too bright or too dim.*
    a. check the contrast and brightness controls
    b. check/troubleshoot/replace the power supply or main PC board

Symptom 10 *You see visible raster scan lines in the display.*
    a. check the contrast and brightness controls
    b. check/troubleshoot/replace the power supply or main PC board
    c. check CRT control voltages
    d. check/replace the FBT
    e. check/replace the horizontal output transistor
    f. replace the main PC board

Symptom 11 *The display flickers or cuts out when the video cable is moved.*
    a. check/tighten the video cable at the video adapter board
    b. check/tighten the video cable connections in the monitor
    c. replace the video cable assembly

Symptom 12 *The image expands in the horizontal direction when the monitor gets warm.*
    a. check/replace capacitors around the HOT circuit
    b. replace the raster board

Symptom 13 *The image shrinks in the horizontal direction when the monitor gets warm.*
    a. check/resolder all soldering connections around the HOT circuit
    b. check/replace the HOT
    c. check/troubleshoot the power supply
    d. replace the power supply (or the entire raster board)

Symptom 14 *High voltage fails after the monitor is warm.*
  a. check/resolder the HOT heat sink connections
  b. check/resolder the FBT connections
  c. try replacing the HOT
  d. try replacing the FBT
  e. check/replace capacitors in the HOT circuit
  f. replace the raster board

Symptom 15 *The image blooms intermittently.*
  a. check/resolder the HOT and FBT areas of the raster board
  b. check video amplifier grounds
  c. try replacing the FBT
  d. replace the raster board
  e. replace the CRT

Symptom 16 *The image appears out of focus.*
  a. try adjusting the focus control (usually on the FBT)
  b. replace the FBT
  c. try rejuvenating the CRT
  d. replace the CRT

Symptom 17 *The image appears to flip or scroll horizontally.*
  a. check/tighten the video cable
  b. try replacing the video cable assembly
  c. check/adjust horizontal hold (if available)
  d. check/resolder connections in the horizontal processing circuit
  e. try replacing the horizontal oscillator IC
  f. replace the raster board

Symptom 18 *The image appears to flip or scroll vertically.*
  a. check/tighten the video cable
  b. try replacing the video cable assembly
  c. check/adjust vertical hold (if available)
  d. check/resolder connections in the vertical processing circuit
  e. try replacing the vertical oscillator IC
  f. replace the raster board

Symptom 19 *The image appears to shake or oscillate in size.*
  a. check the power supply outputs
  b. troubleshoot/replace the power supply if necessary
  c. check/replace capacitors in the HOT circuit
  d. check/resolder HOT and FBT circuit components
  e. replace the raster board

Chapter 8: color monitor troubleshooting

Symptom 1 *The image is saturated with red or appears green-ish-blue (cyan).*
- a. check the red color drive
- b. check the red video amplifier circuit
- c. replace the video drive PC board
- d. check/rejuvenate/replace the CRT

Symptom 2 *The image is saturated with blue or appears yellow.*
- a. check the blue color drive
- b. check the blue video amplifier circuit
- c. replace the video drive PC board
- d. check/rejuvenate/replace the CRT

Symptom 3 *The image is saturated with green or appears bluish-red (magenta).*
- a. check the green color drive
- b. check the green video amplifier circuit
- c. replace the video drive PC board
- d. check/rejuvenate/replace the CRT

Symptom 4 *Raster is present, but there is no image.*
- a. check the monitor installation
- b. check for an interruption in the video drive circuit
- c. check/rejuvenate/replace the CRT
- d. check/replace the video drive PC board

Symptom 5 *A single horizontal line appears in the middle of the display.*
- a. check/replace the vertical oscillator
- b. check/replace faulty vertical output driver transistor(s)
- c. replace the main PC board

Symptom 6 *Only the upper or lower half of an image appears.*
- a. check/replace faulty vertical output driver transistor(s)
- b. replace the main PC board

Symptom 7 *A single vertical line appears along the middle of the display.*
- a. check/replace the horizontal deflection yoke assembly
- b. check/replace the horizontal oscillator
- c. check/replace the horizontal switching transistor(s)
- d. check/replace the FBT
- e. replace the main PC board

Symptom 8 *There is no image and no raster.*
  a. check the monitor's installation
  b. check the contrast and brightness controls
  c. check/troubleshoot/replace the power supply or main PC board
  d. check/replace the horizontal oscillator
  e. check/replace the horizontal output transistor(s)
  f. check/rejuvenate/replace the CRT
  g. check/replace the FBT
  h. replace the main PC board

Symptom 9 *The image is too compressed or too expanded.*
  a. check the front-panel vertical size control(s)
  b. check/replace the vertical oscillator
  c. check for PC problems on the main PC board
  d. replace the main PC board

Symptom 10 *The displayed characters appear to be distorted.*
  a. check the monitor's location and installation
  b. check/correct the image's linearity
  c. check/correct the image's focus
  d. check/replace the FBT
  e. replace the main PC board

Symptom 11 *The display appears wavy.*
  a. check/troubleshoot/replace the power supply or main PC board

Symptom 12 *The display is too bright or too dim.*
  a. check the contrast and brightness controls
  b. check/troubleshoot/replace the power supply or main PC board

Symptom 13 *You see visible raster scan lines in the display.*
  a. check the contrast and brightness controls
  b. check/troubleshoot/replace the power supply or main PC board
  c. check CRT control voltages
  d. check/replace the FBT
  e. check/replace the horizontal output transistor
  f. replace the main PC board

Symptom 14 *Colors bleed or smear.*
  a. check/replace the video cable assembly
  b. check/replace capacitors in the video amplifier circuit

269

    c. check/replace transistors in the video amplifier circuit
    d. replace the video amplifier board

Symptom 15 *Colors appear to change when the monitor is warm.*
    a. check/replace the video cable assembly
    b. check/resolder video board connections
    c. replace the video amplifier board

Symptom 16 *An image appears distorted in 350 or 400 line mode.*
    a. check/adjust screen mode adjustments on the raster board
    b. try replacing the sync sensing circuit

Symptom 17 *The fine detail of high-resolution graphic images appears a bit fuzzy.*
    a. check/correct video amplifier board installation
    b. replace the video amplifier board

Symptom 18 *The display changes color, flickers, or cuts out when the video cable is moved.*
    a. check/tighten the video cable at the video adapter board
    b. check/tighten the video cable connections in the monitor
    c. replace the video cable assembly

Symptom 19 *The image expands in the horizontal direction when the monitor gets warm.*
    a. check/replace capacitors around the HOT circuit
    b. replace the raster board

Symptom 20 *The image shrinks in the horizontal direction when the monitor gets warm.*
    a. check/resolder all soldering connections around the HOT circuit
    b. check/replace the HOT
    c. check/troubleshoot the power supply
    d. replace the power supply (or the entire raster board)

Symptom 21 *High voltage fails after the monitor is warm.*
    a. check/resolder the HOT heat sink connections
    b. check/resolder the FBT connections
    c. try replacing the HOT
    d. try replacing the FBT
    e. check/replace capacitors in the HOT circuit
    f. replace the raster board

Symptom 22 *The image blooms intermittently.*
    a. check/resolder the HOT and FBT areas of the raster board
    b. check video amplifier grounds

c. try replacing the FBT
  d. replace the raster board
  e. replace the CRT

Symptom 23 *The image appears out of focus.*
  a. try adjusting the focus control (usually on the FBT)
  b. replace the FBT
  c. try rejuvenating the CRT
  d. replace the CRT

Symptom 24 *The image appears to flip or scroll horizontally.*
  a. check/tighten the video cable
  b. try replacing the video cable assembly
  c. check/adjust horizontal hold (if available)
  d. check/resolder connections in the horizontal processing circuit
  e. try replacing the horizontal oscillator IC
  f. replace the raster board

Symptom 25 *The image appears to flip or scroll vertically.*
  a. check/tighten the video cable
  b. try replacing the video cable assembly
  c. check/adjust vertical hold (if available)
  d. check/resolder connections in the vertical processing circuit
  e. try replacing the vertical oscillator IC
  f. replace the raster board

Symptom 26 *The image appears to shake or oscillate in size.*
  a. check the power supply outputs
  b. troubleshoot/replace the power supply if necessary
  c. check/replace capacitors in the HOT circuit
  d. check/resolder HOT and FBT circuit components
  e. replace the raster board

Chapter 9: flat-panel displays

Symptom 1 *One or more pixels are defective.*
  a. try rebooting the computer from a cold start
  b. replace the LCD assembly
  c. replace VRAM
  d. replace the motherboard

Symptom 2 *There is poor visibility in the LCD.*
  a. check the LCD contrast control
  b. check/troubleshoot/replace the backlight power supply
  c. check the backlight diffuser
  d. check/replace the CCFTs or EL panel

Symptom 3 *There is poor visibility in the GPD.*
- *a.* check the plasma display brightness control
- *b.* check/troubleshoot/replace the GPD power supply
- *c.* replace the GPD assembly

Symptom 4 *The display is completely dark. There is no apparent display activity.*
- *a.* check for loose connectors/wiring between the system and display
- *b.* check/troubleshoot/replace the backlight power supply
- *c.* check/troubleshoot/replace the system power supply
- *d.* check/replace the display controller IC or motherboard
- *e.* check/replace the display assembly

Symptom 5 *The display appears erratic.*
- *a.* check for loose connectors/wiring between the system and display
- *b.* check/replace the display controller IC
- *c.* check/replace VRAM
- *d.* check/replace the system motherboard
- *e.* check/replace the display assembly

272

Monitor glossary

accelerator a video adapter based on a graphics processor designed to speed up video performance. This is accomplished by relieving graphics processing from the system CPU.

address a unique set of numbers that identifies a particular location in computer memory.

analog monitor a computer monitor that uses continually variable color control voltages. The use of analog voltages allows a large number of colors to be displayed.

anode the positive electrode of a two-terminal semiconductor device or electrical device such as a CRT.

ANSI (American National Standards Institute) an organization that sets standards for electrical and electronic assembly.

architecture a description of how a system is constructed and how its components are put together.

ASCII (American Standard Code for Information Interchange) a set of standard codes defining characters and symbols used by computers.

AVI (Audio-Video Interleave) the native Microsoft format that combines digital images and audio into a composite file that can be distributed or embedded in Windows applications.

bandwidth for a monitor, the number of pixels which can be displayed on the CRT in a 1-s period. Modern high-resolution monitors have bandwidths exceeding 100 MHz (or 100 million pixels per second).

base one of the three leads of a bipolar transistor.

bezel a metal or plastic frame fitting over the LCD glass that holds part of the display system together. Also the front plastic frame that surrounds the face of a CRT.

BIOS (Basic Input/Output System) a series of programs that handle the computer's low-level functions. Video adapters often include their own supplemental BIOS to support high-level video operations.

bit binary digit. The basic unit of digital information written as a 0 or a 1.

bit block transfer also called BitBit. A feature of Windows graphics acceleration that defines a screen image as a series of pixels that can be moved around quickly and easily.

boot the process of initializing a computer and loading a disk operating system.

boot device a drive containing the files and information for a disk operating system.

brightness the amount of light liberated by the CRT measured in foot lumens (ft lm).

buffer a temporary storage place for data.

bus a collection of digital signal lines within a computer. There may be more than one bus within a computer.

byte a set of eight bits. A byte is approximately equivalent to a character.

cache memory simply called "cache." Part of a computer's RAM operating as a buffer between the system RAM and CPU. Recently used data or instructions are stored in cache. RAM is accessed quickly, so data called for again are available right away. This improves overall system performance.

capacitance the measure of a device's ability to store an electric charge. The unit of capacitance is the farad (F).

capacitor an electronic device used to store energy in the form of an electric charge.

cathode the negative electrode of a two-terminal semiconductor device.

CCFT (cold-cathode fluorescent tube) the light source used in edgelighting LCDs so that the display is visible in low light or darkness.

cell gap the space between two pieces of glass that contains the fluid liquid crystal.

CGA (Color Graphics Adapter) a low-resolution graphics mode featuring 320×200 resolution four-color or 640×200 resolution two-color operation; commonly found in pen-computers and older laptop systems.

chip carrier a rectangular or square package with I/O connections on all four sides.

CODEC (COmpressor/DECompressor) an IC typically used to compress analog video signals into digital form during video capture and then decompress those digital files for playback.

collector one of the three leads of a bipolar transistor.

color display a monitor capable of displaying images in color using three primary color signals (red, green, and blue—sometimes referred to as an RGB monitor). If 64 levels of each primary color can be generated, the monitor can produce $64 \times 64 \times 64$ or 262,144 colors.

configuration the components that make up a computer's hardware setup.

contrast ratio the difference in luminance between a selected (on) pixel and an unselected (off) pixel.

convergence in a color CRT, the alignment of electron beams such that each will come together at an aperture in the CRT's shadow mask. See *shadow mask*.

CPU (central processing unit) also called a microprocessor. The primary functioning unit of a computer system.

CRT (cathode-ray tube) a vacuum tube with a wide face coated with a phosphor that generates light when bombarded with a focused electron beam. CRTs are the foundation for most PC monitors in service today.

CRTC (CRT controller) an IC used in basic video adapters to form the interface between video memory and the monitor.

deflection yoke a tightly wound set of coils placed around the CRT neck and used to direct the electron beam around the CRT. Often designated H-DY for the horizontal deflection yoke, and V-DY for the vertical deflection yoke.

degaussing coil a coil of wire powered with ac which is used to clear magnetized areas of the CRT.

dichroic host also known as guest host. A type of liquid crystal fluid in which color dye is added.

DMA (direct memory access) a fast method of moving data from a storage device directly to RAM.

DOS (disk operating system) a program or set of programs that directs the operations of a disk-based computing system.

DOS extender software that uses the capabilities of advanced microprocessors running under DOS to access more than 640 kbytes of RAM.

275

dot   in a color CRT, one of the three primary color phosphors (red, green, or blue). Dots are arranged in a close triangle so that they appear as a single point to the unaided eye.

DPMS (Display Power Management System)   a recent standard employed for monitor power conservation.

drain   one of the three leads of a field-effect transistor.

driver   also called a device driver. A small segment of software that allows application software to take advantage of specialized hardware such as the advanced video modes of a video adapter.

dynamic mode switching   a system's ability to toggle between several resolutions or color depths in Windows without having to restart Windows.

EGA (enhanced graphics adapter)   a medium-resolution graphics mode featuring up to 640×350 resolution, 16-color operation; commonly available in older notebook systems.

EIA (Electronics Industry Association)   a standards organization in the United States that develops specifications for interface equipment.

EIAJ   a Japanese standards organization which is the equivalent of the US Joint Electronic Device Engineering Council (JEDEC).

EL (electroluminescent)   material that glows when voltage is applied across it. Many monochrome LCDs use a sheet of EL material as a backlight so the LCD can be seen in darkness.

emitter   one of the three leads of a bipolar transistor.

EMS (extended memory system)   a highly integrated IC controller used to access extra RAM.

ESD (electrostatic discharge)   the sudden accidental release of electrons accumulated in the body or inanimate objects. Static charges are destructive to MOS ICs and other semiconductors.

expansion slot connector   a bus that connects an expansion board to the PC. There are several different types of busses: ISA, VL, and PCI.

FBT (flyback transformer)   a high-energy transformer used to convert horizontal pulse signals into high-voltage that drives the CRT anode as well as other voltages used within the monitor.

file   a collection of related information which is stored together on disk.

filter   part of a power supply circuit intended to smooth the ripple of pulsating DC produced by the rectifier.

firmware program instructions recorded on a permanent memory device such as a PROM or EPROM. BIOS is a typical example of firmware.

fixed-frequency monitor an analog monitor that operates using a fixed horizontal sync frequency. Vertical sync frequency may change slightly to accommodate different screen modes.

flatpack one of the oldest surface-mount packages with 14 to 50 ribbon leads on both sides of its body.

focus electrode a CRT electrode that focuses the electron beam(s).

FSTN (film-compensated STN) a modified version of super twist nematic liquid crystal material.

gate one of the three leads of a field-effect transistor.

ghosting a phenomenon where voltage from an energized element leaks to an adjacent off element, which seems to turn the off element partially on. An effect typically found in passive matrix LCDs.

GPD (gas plasma display) a flat-panel display design that forms monochrome images by ionizing discrete points of neon/xenon gas.

GUI (graphic user interface) software that allows a user to interact with a program.

Hercules Graphics a video mode that provides bitmapped single-color graphics at a resolution of 720×348.

high memory the RAM locations residing between 640 kbytes and 1 Mbyte.

horizontal sync the horizontal synchronization signal used to time each row of pixels.

IDE (Integrated Drive Electronics) a physical interface standard commonly used in medium to large hard drives. IDE control electronics are housed in the drive itself instead of an external control board.

inductance the measure of a device's ability to store a magnetic charge. The unit of inductance is the henry (H).

inductor an electronic device used to store energy in the form of a magnetic charge.

interface a hardware-software connection that links one device to another.

interlace an image displayed by drawing two passes of the electron beam where each pass strikes every other horizontal line.

JEDEC (Joint Electronic Device Engineering Council) the U.S. standards organization that handles packaging and assembly standards.

JEIDA (Japanese Electronics Industry Development Association) the Japanese equivalent of JEDEC dedicated to Asian electronics packaging and assembly.

LCD (liquid crystal display) a display technology using a thin layer of liquid crystal sandwiched between two electrodes. An electric field across the liquid crystal causes the crystals to rotate and appear opaque.

LIF (low insertion force) sockets that require only a minimum force to insert or extract an IC.

linearity the evenness of a monitor display which keeps circles round and squares even.

local bus an interface bus architecture that provides a faster, more efficient path to the CPU. Devices using a local bus interface are intended to achieve higher performance than conventional ISA bus devices.

logic analyzer an instrument used to monitor signals of an integrated circuit or system.

mapping the area in computer memory where a particular video mode stores its data. Different video modes store data in different locations.

MCGA (multicolor graphics array) a video mode which provides CGA support with $640 \times 480 \times 2$ and $320 \times 200 \times 256$ video modes.

MCP (math coprocessor) a sophisticated processing IC intended to enhance the processing of a computer by performing floating-point math operations instead of the CPU.

MDA (monochrome display adapter) the first video standard defined for the IBM PC.

monochrome monitor a computer monitor that displays images in a single color.

motherboard also called the main logic board. In a small computer, the major PC board containing the CPU, core memory, and most of the system's controller ICs.

multifrequency monitor a monitor that is designed with variable horizontal and vertical sync frequencies. This allows the multifrequency monitor to support a large number of video modes.

OS (operating system)   the interface between the hardware and software running on your PC. Typical operating systems for the PC are DOS, Windows 3.1, and Windows 95.

page   a reference to a block of memory (often video memory) in a computer.

palette   a range of colors that is available to be displayed. Only a certain number of the possible colors that can be produced are available at a given time. For example, a monitor that is capable of producing 262,144 colors may only be able to show 256 of those colors at any one time.

parallel port   a physical connection on a computer used to connect output devices. Data are transmitted as multiple bits sent together over separate wires. Typically used to connect a printer.

parity   a means of error checking using an extra bit added to each transmitted character.

permeable   the ability of a material to be magnetized.

persistence   the duration which an illuminated phosphor will remain lit in the absence of an electron beam.

pincushion   a form of visual distortion that occurs when a flat image is projected onto a curved surface. Monitor raster circuits are designed to compensate for this distortion.

pitch   also called dot pitch. The center-to-center dimension between adjacent pixels.

pixel (picture element)   a single visual element of a computer display. On a color CRT, a pixel is composed of three dots (red, green, and blue).

polarizers   sheet material made of polymer acetate incorporated with iodide molecules. The molecules allow scattered light to enter in one plane only. TN LCDs require two polarizers, one in front and one in back.

POST (power-on self-test)   a program in BIOS which handles the computer's initialization and self-test before loading DOS.

QWERTY   a standard typewriter-style keyboard. The top row of letters begins Q, W, E, R, T, Y, U, I, etc.

raster   the overall sweep of the electron beam across a CRT which occurs regularly (even in the absence of data).

rectifier   a device (usually a semiconductor diode) that converts ac to pulsating dc by cutting off one polarity of the ac signal. The basis of all power supplies.

279

regulation the ability of a power supply to maintain a constant output voltage as load conditions change.

resistance the measure of a device's opposition to the flow of current. The unit of resistance is the ohm (Ω).

resistor a device used to limit the flow of current in an electronic circuit.

resolution the number of individual pixels which comprise a complete display frame (e.g., 320×200 or 640×480).

retrace the blanking of a video signal while the electron beam returns to the start of a new horizontal line or page.

sawtooth wave an electrical signal resembling the tooth of a saw—a long sloping leading edge followed by an almost vertical drop. Typically found in a monitor's vertical drive circuit.

screen electrode a CRT electrode used to regulate overall display brightness.

SCSI (Small-Computer System Interface) a physical interface standard for large to huge (up to 3 Gbytes) hard drives.

serial port a physical connection on a computer used to connect output devices. Data are transmitted as individual bits sent one at a time over a single wire. Typically used to connect a modem or mouse.

shadow mask also known as a slot mask in some CRT types. A thin metal plate (placed directly in front of a color CRT's phosphor face) which blocks stray electrons from exciting nearby color dots. A slot mask is a variation of the shadow mask.

SIMM (single in-line memory module) a quantity of RAM mounted onto a convenient add-on module that can be plugged into PC motherboards or some video boards.

SMT surface-mount technology.

source one of the three leads of a field-effect transistor.

ST (super twist) also called STN for super twisted nematic. An improved liquid crystal material which is more stable and provides much better contrast than regular twisted nematic liquid crystal.

synchronous circuit operation where signals are coordinated through the use of a master clock.

TFT (thin-film transistor) a fabrication technology used in flat-panel displays where transistors used to operate each pixel are fabricated right onto flat-panel displays.

TN (twisted nematic) a type of liquid crystal material.

transfer rate the speed at which a hard or floppy drive can transfer information between its media and the CPU, typically measured in Mbits per second.

TSR (Terminate and Stay Resident) a program residing in memory that can be invoked from other application programs.

TST (triple super twist) an improved liquid crystal material which provides better contrast than regular super twist liquid crystal.

TTL (transistor-transistor logic) digital logic ICs using bipolar transistors.

vertical sync a vertical synchronization signal used to time each frame (set of horizontal lines).

VESA (Video Electronics Standards Association) an industry-driven coalition of video and monitor makers dedicated to developing and implementing standards for video systems and monitors.

VGA (video graphics array) a high-resolution graphics mode featuring 640×480 16-color and 320×200 256-color resolutions; widely available in pen-computers and notebook systems.

video memory an amount of RAM used to store image data to be displayed by a monitor.

viewing angle a conceptual cone perpendicular to the LCD where the display can still be seen clearly.

VRAM (Video RAM) specialized memory used in video accelerator boards. Dual data busses in the VRAM allow new data to be written at the same time data are being read, so apparent video performance is improved.

C

Index of resources

Monitor manufacturers

Acer America Corp.
2641 Orchard Parkway
San Jose, CA 95134
Tel: 800-733-2237

Amdek
Division of Wyse
3471 N. First Street
San Jose, CA 95134
Tel: 408-922-5700

American Mitac Corp.
410 E. Plimeria Drive
San Jose, CA 95134
Tel: 800-648-2287

AST Research
16215 Alton Parkway
Irvine, CA 92713
Tel: 714-727-4141

ATI Technologies
3761 Victoria Park Avenue
Scarborough, Ontario M1W 3S2
Canada
Tel: 416-756-0718

CTX International, Inc.
20530 Earlgate Street
Walnut, CA 91789
Tel: 714-595-6146

Diamond Computer Systems
470 Lakeside Drive
Sunnyvale, CA 94086
Tel: 408-736-2000

Epson America, Inc.
20770 Madrona Avenue
Torrance, CA 90503
Tel: 800-922-8911

Everex Systems
48431 Milmont Drive
Fremont, CA 94538
Tel: 800-821-0806

GoldStar Technologies
3003 N. First Street
San Jose, CA 95134-2004
Tel: 800-777-1194

Hewlett-Packard Company
California Personal Computer Division
974 E. Arques Avenue
P.O. Box 3486
Sunnyvale, CA 94086
Tel: 800-752-0900

Idek/Iiyama North America, Inc.
650 Louis Drive, #120
Warminster, PA 18974
Tel: 215-957-6543

Leading Technology
10430 S.W. Fifth Street
Beaverton, OR 97005
Tel: 503-646-3424

MAG Innovision, Inc.
4392 Corporate Center Drive
Los Alamitos, CA 90720
Tel: 800-827-3998

Magnavox
Phillips Consumer Electronics
One Phillips Drive
Knoxville, TN 37914
Tel: 615-521-4316

Mitsubishi Electronics America, Inc.
Information Systems Division
5757 Plaza Drive
P.O. Box 6007
Cypress, CA 90630
Tel: 800-843-2515

Nanao USA Corp.
23535 Telo Avenue
Torrance, CA 90505
Tel: 800-800-5202

NEC Technologies, Inc.
1255 Michael Drive
Wood Dale, IL 60191
Tel: 800-388-8888

Packard Bell Electronics
9425 Canoga Avenue
Chatsworth, CA 91311
Tel: 818-773-4400

Panasonic
2 Panasonic Way
Secaucus, NJ 07094
Tel: 800-346-4768

Princeton Graphic Systems
1100 N. Meadow Parkway, #150
Roswell, GA 30076
Tel: 800-221-1490

Relisys
320 S. Milpitas Boulevard
Milpitas, CA 95035
Tel: 408-945-9000

Sampo Corp. of America
5550 Peachtree Industrial Boulevard
Norcross, GA 30071
Tel: 404-449-6220

Samsung Electronics America
3655 N. First Street
San Jose, CA 95134
Tel: 800-446-0262

Seiko Instruments USA, Inc.
Color Graphics Group
1130 Ringwood Court
San Jose, CA 95131
Tel: 800-888-0817

Sony Computer Peripheral Products Co.
655 River Oaks Parkway
San Jose, CA 95134
Tel: 800-352-7669

Tatung Co. of America, Inc.
2850 El Presidio Street
Long Beach, CA 90810
Tel: 800-829-2850

Toshiba America Consumer Products, Inc.
1010 Johnson Drive
Buffalo Grove, IL 60089-6900
Tel: 708-541-9400

ViewSonic
12130 Mora Drive
Santa Fe Springs, CA 90670
Tel: 800-888-8583

Video adapter/accelerator board manufacturers

Actix Systems, Inc.
3060 Tasman Drive
Santa Clara, CA 95054

Tel: 800-927-5557
Tel: 408-986-1625
Fax: 408-986-1646

Advanced Integration Research, Inc.
(AIR)
2188 Del Franco Street
San Jose, CA 95131
Tel: 800-866-1945
Tel: 408-428-0800
Fax: 408-428-0950

American Megatrends, Inc. (AMI)
6145-F Northbelt Parkway
Norcross, GA 30071
Tel: 800-828-9264
Tel: 404-263-8181
Fax: 404-263-9381

Appian Technology, Inc.
477 N. Mathilda Avenue
Sunnyvale, CA 94086
Tel: 800-727-7426
Tel: 408-730-5400
Fax: 408-730-5473

ATI Technologies, Inc.
33 Commerce Valley Drive East
Thornhill, Ontario L3T 7N6
Canada
Tel: 416-882-2600
Fax: 416-882-2620

Boca Research, Inc.
6413 Congress Road
Boca Raton, FL 33487
Tel: 407-997-6227
Fax: 407-997-0918

Cardinal Technologies, Inc.
1827 Freedom Road
Lancaster, PA 17601
Tel: 717-293-3000
Fax: 717-293-3055

Diamond Computer Systems, Inc.
1130 E. Arques Avenue
Sunnyvale, CA 94086
Tel: 408-736-2000
Fax: 408-730-5750

Elitegroup Computer Systems, Inc.
45225 Northport Court
Fremont, CA 94538
Tel: 800-829-8890
Tel: 510-226-7333
Fax: 510-226-7350

ELSA, Inc.
400 Oyster Point Boulevard, #109S
San Francisco, CA 94080
Tel: 800-272-ELSA
Tel: 415-615-7799
Fax: 415-588-0113

Focus Information Systems, Inc.
4046 Clipper Court
Fremont, CA 94538
Tel: 800-925-2378
Tel: 510-657-2845
Fax: 510-657-4158

Genoa Systems Corporation
75 East Trimble Road
San Jose, CA 95131
Tel: 800-934-3662
Tel: 408-432-9090
Fax: 408-434-0997

Hercules Computer Technology, Inc.
3839 Spinnaker Court
Fremont, CA 94538
Tel: 800-532-0600
Tel: 510-623-6030

Matrox Electronic Systems Ltd.
1055 St. Regis Boulevard
Dorval, Quebec H9P 2T4
Canada
Tel: 514-685-2630
Fax: 514-685-2853

Metheus Corporation
1600 N.W. Compton Drive
Beaverton, OR 97006
Tel: 800-638-4387
Tel: 503-690-1550
Fax: 503-690-1525

284

Micro-Labs, Inc.
204 Lost Canyon Court
Richardson, TX 75080
Tel: 214-234-5842
Fax: 214-234-5896

Nth Graphics Ltd.
11500 Metric Boulevard, #210
Austin, TX 78758
Tel: 800-624-7552
Tel: 512-832-1944
Fax: 512-832-5954

Number Nine Computer Corporation
18 Hartwell Avenue
Lexington, MA 21731
Tel: 800-GET-NINE
Tel: 617-674-0009
Fax: 617-674-2919

Orchid Technology, Inc.
45365 Northport Loop W.
Fremont, CA 94538
Tel: 800-767-2443
Tel: 510-683-0300
Fax: 510-490-9312

Sigma Designs, Inc.
47900 Bayside Parkway
Fremont, CA 94538
Tel: 800-845-8086
Tel: 510-770-0100
Fax: 510-770-2640

SPIDER Graphics, Inc.
801 Ames Avenue
Milpitas, CA 95035
Tel: 408-956-1231
Fax: 408-956-1342

VidTech Microsystems, Inc.
1700 93rd Lane, N.E.
Minneapolis, MN 55449
Tel: 800-752-8033
Tel: 612-780-8033
Fax: 612-780-2040

Volante, Inc.
1515 Capital of Texas Highway S.
5th Floor
Austin, TX 78746
Tel: 800-253-8831
Tel: 512-329-5055
Fax: 512-329-6326

Western Digital Corporation
8105 Irvine Center Drive
Irvine, CA 92718
Tel: 800-832-4778
Tel: 714-932-4900
Fax: 714-932-6498

Monitor service providers

The following listing of independent monitor service houses is provided for your reference. The vendors listed here are not endorsed by or affiliated with the author or publisher. Readers are advised to request references, prices, delivery, and any other pertinent information before doing business with third- and fourth-party service organizations.

3E Corporation
165 Front Street
Chicopee, MA 01013

Tel: 413-594-2772
Fax: 413-594-7283

CNS (Computer Network Services)
100 Ford Road
Denville, NJ 07834
Tel: 201-625-4056
Fax: 201-625-9489

CSE, Inc.
780 W. Belden
Addison, IL 60101
Tel: 708-628-7000 x7283

Diversified Electronic Services
4741 Troudsdale Drive
Suite 8
Nashville, TN 37220
Tel: 800-737-9920

Electroservice Laboratories
6085 Sikorsky Street
Ventura, CA 93003
Tel: 805-644-2944
Fax: 805-644-5006

Fessenden Technologies
116 N. 3rd Street
Ozark, MO 65721
Tel: 417-485-2501
Fax: 417-485-3133

Hong Video Technology, Inc.
4467 Park Drive N.W., Suite E
Norcross, GA 30093
Tel: 404-931-0346
Fax: 404-279-8520

Impact
10435 Burnet Road
Suite 114
Austin, TX 78758
Tel: 512-832-9151
Fax: 512-832-9321

Kennsco Component Services
2500 Broadway Street N.E.
Minneapolis, MN 55413
Tel: 800-525-5608
Fax: 717-258-1952

Metro Tech Center
12001 N. Tejon, Suite 110
Westminster, CO 80234
Tel: 800-227-3504

Northstar
7940 Ranchers Road
Minneapolis, MN 55432
Tel: 800-969-0009
Fax: 612-785-1135

Parts, materials, and test instruments

The following listing of independent parts, materials, and instrument vendors is provided for your reference. The vendors listed
here are not endorsed by or affiliated with the author or publisher.
Readers are advised to request a catalog/references, prices, delivery, and any other pertinent information before doing business with
mail order organizations.

Anatek Corp.
RESOLVE Database
P.O. Box 1200
Amherst, NH 03031-1200
Tel: 603-673-4342

B+K Precision
6470 W. Cortland Street
Chicago, IL 60635
Tel: 312-889-1448

Computer Component Source, Inc.
135 Eileen Way
Syosset, NY 11791
Tel: 516-496-8780
Fax: 516-496-8984

Consolidated Electronics
705 Watervliet Avenue
Dayton, OH 45420-2599
Tel: 800-543-3568
Fax: 513-252-4066

Dalbani Corporation
2733 Carrier Avenue
Los Angeles, CA 90049
Tel: 213-727-0054
Fax: 213-727-6032

Digi-Key Corporation
701 Brooks Avenue S.
P.O. Box 677
Thief River Falls, MN 56701-0677
Tel: 800-344-4539
Fax: 215-681-3380

Howard W. Sams & Company
2647 Waterfront Parkway East Drive
Indianapolis, IN 46214-2041
Tel: 800-428-7267

IVS
IMPACT Database
13311 Stark Rd.
Livonia, MI 48150
Tel: 313-261-8801

Jensen Tools, Inc.
7815 S. 46th Street
Phoenix, AZ 85044-5399
Tel: 800-426-1194
Fax: 800-366-9662

MCM Electronics
650 Congress Park Drive
Centerville, OH 45459-4072
Tel: 800-543-4330
Fax: 513-434-6959

Parts Express International, Inc.
340 E. First Street
Dayton, OH 45402-1257
Tel: 513-222-1073
Fax: 513-222-4644

Premium Parts
P.O. Box 28
Whitewater, WI 53190-0028
Tel: 800-558-9572
Fax: 414-473-4727

Print Products International
8931 Brookville Road
Silver Spring, MD 20910
Tel: 800-638-2020
Fax: 800-545-0058

Index

Illustrations are indicated in **boldface.**

290

291

292

293

295

Switching power supplies,
troubleshooting,
continued
fuses blowing, 161–162,
264
intermittent operation,
160–161, 264
monitor dead/no
raster/no picture,
158–160, 263–264
Switching regulation,
concepts, 153–155
Synchronization, 12, 17–18
composite, 18
horizontal, 7, 17, 32, 41
separate, 18
vertical, 7, 18, 114

T

Temperature, monitors and,
12–13
Terminate-and-stay resident
(TSR), 116
Test equipment, 61–91 (*see
also* Tools)
circuit analyzer, 77–78
computer-aided, 90–91
CRT testers-restorers,
87–88, **88**
high-voltage probes,
86–87, **86**
logic probe, 79–80
monitor testers-analyzers,
89–90, **89**, **90**
multimeters, 64–77
oscilloscopes, 80–86
suppliers of, 286–287
Tests and testing, 122–138
blank raster, 258
burn-in, 176, 258–259
capacitance, 70–73, **72**
color bars, 252
color drive, 137–138
color purity, 135–137, **136**
convergence, 126, 127,
252–255
current, 68–69, **70**
diodes, 73–75, **74**
dynamic convergence,
133–134, **135**
dynamic pincushion,
125–126, **126**

Tests and testing, *continued*
focus, 124–125, 256–257,
125
frequency, 85–86
gray scale linearity, 252
high-voltage, 122–124, 258
horizontal centering,
127–128
horizontal linearity,
129–130, **130**
horizontal phase, 126–127,
127
horizontal size, 128–129
integrated circuits (ICs),
77
linearity, 255–256, **256**
phase test, 256
purity, 257
resistance, 69–70, **71**
screen control, 124
static convergence,
131–133, **132**, **133**
time, 85–86
transistors, 75–77, **76**
using MONITORS utility
for diagnostics,
244–260
vertical centering, 127–128
vertical linearity, 129–130,
130
vertical size, 128–129
video adapters, 249–250
voltage, 67–68, 83–85, **68**
vs. alignment, 117–118
white purity, 125, 136, 137
Text mode, 94
TFT, 230–231, 233
Thermistors, 152, 162
Thin-film transistor (TFT),
230–231, 233
Time, measuring with
oscilloscopes, 85–86
Tools (*see also* Test
equipment)
diagonal cutters, 63
needlenose pliers, 63
screwdrivers, 62
soldering, 63–64
tweezers, 63
wrenches, 63
Transflective LCD viewing
mode, 226

Transformers, 140–141
flyback (FBT), 19–20, 33,
42–43, 195–196, **34**
isolation, 54
step-down, 141
Transistors
checking with multimeter,
75–77, **76**
driver, 231
horizontal output, 173
thin-film (TFT), 230–231,
233
Transistor-transistor logic
(TTL), 92
Transmissive LCD viewing
mode, 226
Triads, 3–4, 37
Troubleshooting
backlight appears
inoperative, 164–165,
264
backlight power supplies,
164–165, 264
cathode-ray tube (CRT),
58–60
color monitors, 188–213,
268–271
colors bleed or smear, 208,
269–270
colors change when
monitor warm,
208–209, 270
computer on/no display,
113, 262
computer sounds one long
beep/no display,
113–114, 262
computer sounds two
short beeps/no
display, 113–114, 262
computer-aided, 90–91
display changes color, 210,
270
display cuts out, 210, 270
display flickers, 183, 210,
266, 270
display image rolls,
114–115, 262
display is dark/no activity,
241–242, 272
display too bright or too
dim, 182, 266, 269

About the Author

Stephen J. Bigelow has written a number of best-selling books, including *Troubleshooting and Repairing PC Drives and Memory Systems, Troubleshooting and Repairing Notebook, Palmtop, and Pen Computers,* and *Troubleshooting, Maintaining, and Repairing Personal Computers: A Technician's Guide.* He has also written scores of articles for such highly regarded electronics magazines as *Popular Electronics, Electronics NOW,* and *Electronic Service & Technology.* Bigelow is editor and publisher of *The PC Toolbox,* a top PC service newsletter for computer enthusiasts and technicians. He is also founder and president of Dynamic Learning Systems, a writing, research, and publishing company specializing in electronics and PC service topics.

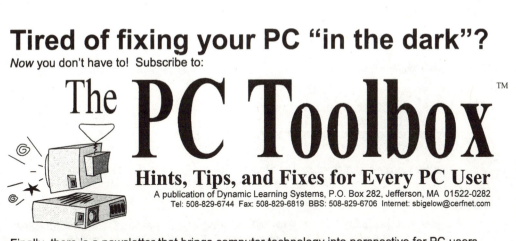

The PC Toolbox™/MONITORS

Use this form when ordering *The PC Toolbox*™ or the registered version of **MONITORS**.
You may tear out or photocopy this order form.

YES! Please accept my order as shown below: (check any one)

_____ Send me the registered version of **MONITORS** for **$20** (US)
Massachusetts residents please add $1 sales tax.

Keep the software, but start my 1 year subscription (6 issues) to *The PC Toolbox*™ for
_____ **$39** (US). I understand that I have an unconditional 90 day money-back guarantee
with the newsletter.

**A special offer for readers of "Troubleshooting and Repairing Computer Monitors
(2nd ed)"!** I'll take the registered version of **MONITORS** *and* the 1 year subscription to
The PC Toolbox™ (6 issues) for *only* **$49** (US). I understand that the newsletter has an
unconditional 90 day money-back guarantee.

SPECIFY YOUR DISK SIZE: (check any one)

_____ **3.5"** High-Density (1.44MB)   _____ **5.25"** High-Density (1.2MB)

PRINT YOUR MAILING INFORMATION HERE:

Name: Company:

Address:

City, State, Zip:

Country:

Telephone: (   )   Fax: (   )

PLACING YOUR ORDER:

By FAX: Fax this completed order form (24 hrs/day, 7 days/week) to **508-829-6819**

By Phone: Phone in your order (Mon-Fri; 9am-4pm EST) to **508-829-6744**

___ MasterCard   Card: ___ ___ ___ ___ ___ ___ ___ ___ ___ ___ ___ ___ ___ ___ ___ ___

___ VISA   Exp: ___/___   Sig: _____

Or by Mail: Mail this completed form, along with your check, money order, PO, or credit card info to:
Dynamic Learning Systems, P.O. Box 282, Jefferson, MA 01522-0282 USA